ELECTRI L
APPLIANCES
TWENTIETH-CENTURY DESIGN

Penny Sparke

E.P. DUTTON New York

First published, 1987, in the United States
by E.P. Dutton

Published in the United States by E.P.
Dutton, a division of New American
Library,
2 Park Avenue,
New York, N.Y. 10016.

Library of Congress Catalog Card Number:
86-73137

ISBN:0-525-24532-4 (Cloth)
ISBN:0-525-48298-9 (DP)

OBE

10 9 8 7 6 5 4 3 2 1
First Edition

Designed by Richard Crawford

Typeset by TJB Photosetting Ltd, South
Witham, Lincolnshire, England.
Produced in Great Britain.

*Title page: Electric kettle made by Bulpitt & Sons in 1922. It was the first
electric kettle with the immersion element inside.*

CONTENTS

FRIGIDAIRE Frost-Proof
Refrigerator-Freezer... **Another Frigidaire "First"**-
Freezing Without Frosting! No frost ever
in refrigerator section or in the freezer.

Model FPD-130-59

Frigidaire's 'Frost-Proof' refrigerator-freezer, 1959.

INTRODUCTION

"Labour saved is leisure gained."

ALTHOUGH a number of different energy sources have been used throughout this century to power utensils in the home, none of them has dominated to the same extent as electricity. Since the discovery of electric current as a source of light, heat, and motive power, our domestic environments have been transformed by the presence of a multitude of powered tools which were invented, or so we were told, with the sole purpose of making our lives easier. This book tells the story of the way in which these appliances have developed through this century, and how and why they have become such an important factor within modern everyday life.

The years 1880–1980 saw the emergence and development of countless electrical appliances used, predominantly, by women in the

A CLOCK THAT MAKES TEA!

Calls the sleeper at a given hour, automatically lights spirit lamp, boils a pint of water for tea, shaving, and other purposes, pours it into a pot, extinguishes lamp, and finally rings second bell to signify all is ready. Invaluable to Ladies, Nurses, Professional and Business Men. It is strong, simple, and in no sense a toy. Prices 25s. to 70s. Postage in United Kingdom 1s. extra. With Foreign orders sufficient postage to cover 11 lb. Weight should be sent.

Please send for Illustrated Booklet, post free from

AUTOMATIC WATER BOILER CO.,
26a, Corporation St., Birmingham,
LONDON OFFICE AND SHOWROOM—
31, George Street, Hanover Square.

home. Any account of these objects, which aims to do more than merely identify them, must consider both their demand and their supply. This entails looking at the social, economic, technological and cultural reasons why these goods were desired and purchased, as well as examining the commercial base of their manufacture. In turn, such a broad context must be supplemented by a discussion of the more abstract constraints upon the nature of their development — i.e. of the ideological and human factors that have shaped them.

The first half of this book sets out, therefore, to cover the broad ground, albeit in a necessarily condensed fashion. It aims to provide a setting and rationale for a more detailed account of the artefacts in question which will

Automatic tea-making machine advertisement from 'Madam', 1904.

form the second half of this study.

If a single theme dominates, it is the rôle that domestic appliances have played in communicating a particular set of values to their users who are, for the most part, women. Whether or not women actually paid for their appliances, the consumption choices were invariably theirs. Those decisions were always made, however, under the influence of intense marketing and advertising programmes and in a situation of restricted choice. The number of, say, washing-machines, available on the market at any one time has always been determined by the number of manufacturers producing them and, from the 1930s onwards, most models have been available for relatively short periods of time only. Also, unlike furniture, few people buy second-hand domestic appliances. As a result, the domestic appliance manufacturer has had a very strong control over the market. The cultural values embedded within the products have been those of the system of mass-production within a capitalistic economy and society.

The general picture of appliance consumption in this century has been one of a rapidly expanding market. Continual efforts have been made by manufacturers to penetrate new markets by producing an increasing number of cheaper and cheaper goods. This policy has brought with

First advertisement for Belling electric fires, 1912.

it the need for aggressive marketing and finding ways of making appliances more attractive. Therefore, while technological development has facilitated volume production and decreased prices, the need to make products desirable to new markets has placed a growing emphasis on design.

The manufacturers' means of communication has been through the embodiment of certain social myths – among them rationalism, progress, efficiency and hygiene. These were all developed during the era of American industrialization at the end of the

19th century when the appliance industry first developed. The American appliance manufacturers, such as the electrical engineering firms General Electric and Westinghouse, and the divisions of American automobile companies – Frigidaire (General Motors) and Kelvinator (American Motors) – are among the giant companies of this century. Together they are responsible for the most sophisticated developments in mass-production and mass-marketing techniques. It is not surprising therefore, that appliances have become such an important element within the development of mass-consumption. Nor is it surprising that the values associated with an increase of consumption have been so clearly expressed by the changing forms of electrical appliances.

While appliances have both supported and sustained the economic *status quo* of the modern industrialized world, they have also helped to reinforce its social structure, in particular the dominance of the nuclear family, which is part and parcel of that system. In *The Grand Domestic Revolution*, the author Dolores Hayden has, for example, described the way in which a group of women, associated with the Liberation Movement in the 1920s, warned against the consumption of those domestic appliances which did not meet real needs. The movement proposed instead the collective purchasing of only those items which they felt did reduce labour. Advertisers were quick to cash in on the Liberation Movement claiming, for instance, that electric toasters made women 'free'. In the 1930s, a number of American manufacturers paid Lillian Gilbreth and Christine Frederick, two of the leading lights of the women's 'home efficiency' movement (a movement which feminists have since disclaimed as a help to their cause) to help convince housewives of the advantages of their products. In 1929, Mrs Frederick wrote a book on the subject, entitled *Selling Mrs Consumer*, in which she set out to persuade women that appliances could improve their lives.

Appliances have provided a number of feminist writers with a focus for their anger and frustration. In their books, *Never Done: A History of American Housework* and *More Work for Mother*, for instance, the American feminist historians Susan Strasser and Ruth Cowan Schwartz have pointed out that the concept of 'labour-saving' did not necessarily guarantee that the time saved was not going to be spent on other forms of housework. In fact, statistics from the USA in the 1920s, suggest that just as much time in total was spent on housework — while many tasks were removed from the housewife, others, such as shopping,

How thrilled you'll be with an

HMV

SUPER ALL-PURPOSE IRON

IN 4 GAY COLOURS

Brochure for HMV's standard IC10 iron of the 1950s.

replaced them.

These writers have also questioned the process that turned overburdened housewives into alienated consumers. Their suggestions are part of the broad discussion about domestic appliances and cannot be ignored. While they are beyond the immediate scope of this book they are, nonetheless, present by implication. The predominance, for instance, of the home washing-machine over the commercial laundry in this century is a prime example of the way commercial interests have controlled social ideals. A number of failed projects, such as commercial vacuum-cleaning and various co-operative ventures, also bear witness to the power of the corporations to dictate the nature of the products available on the market. Ruth Cowan Schwartz has argued that the gas refrigerator might have had more of a chance if the General Electric Company had not been so determined to lead the market with its electric version. Similarly, the powerful Maytag Washing Machine Company succeeded in suppressing nearly all its competitors in the inter-War years, leaving us with only one technique of washing clothes electrically.

The five chapters making up the first half of this book are of a general nature, focusing on themes which illuminate the broad history of modern domestic appliances. The first chapter examines the question of the demand for appliances after 1880 in relation to two frequently voiced hypotheses — that the demand for domestic appliances expanded as a direct result of the demise of the household servant, and that there is a direct correlation between the increased use of appliances and the changing rôle of women in this century. Chapter Two looks at the way different approaches towards the kitchen have influenced the nature of the appliances within it and vice versa. Chapter Three swings the pendulum back from the area of demand to that of supply. It examines the nature of the domestic appliance industry, in the USA and elsewhere, showing its commitment, from the early days, to high volume production. The fourth chapter focuses on the important question of new technologies in the general area of electrical domestic appliances, emphasizing their essential link with the need to find ways of producing more goods at low prices. Finally, the fifth chapter deals with the question of appliance design and the relationship, since the late 1920s, of the industrial designer with electrical goods. It stresses the way in which this development has, essentially, been a strategic one, making goods increasingly desirable to an ever widening market, and encouraging consumption through accelerated object obsolescence.

In conjunction with each other, these themes form the necessary backcloth against which the stories of particular appliances can be told.

Brochure for Hoover 'Spinarinse' - early 1960s.

1

THE 'SERVANT PROBLEM' AND THE 'NEW WOMAN'

"The mechanization of the household had its starting point in social problems: the status of American women and the status of domestic servants."

S. Giedion *Mechanization Takes Command*
1948

TWO important social changes which have taken place during the period 1880 to 1980 have directly affected the mechanization of the domestic work area – the disappearance of the servant and the changing rôle of the housewife. The first transformation had its greatest impact within the middle-class sector of industrialized society, while the second went on to affect society as a whole. Both can be seen as important forces behind the impulse to discover new ways of minimizing labour in the home, and developing domestic machines to replace the work of the hand.

The 'servant problem' was described as such by the American middle-class 'mistresses' who, from the second half of the 19th century, found that there was a decreasing supply of domestic labourers to fulfil their needs. The 'problem' influenced the influx of labour-saving devices into the home before and after the First World War, but its impact was less significant in subsequent decades as the live-in servant had all but disappeared by that time from middle-class houses.

The question of domestic servants was most hotly discussed in the last decades of the 19th century and the early decades of the 20th century, particularly in the USA. The issue infiltrated numerous women's magazines and caused such organizations as The Society for the Encouragement of Faithful Domestic Servants to be formed in New York in an attempt to fill the growing demand for, and supplement the diminishing supplies of, domestic workers. The problem had a special significance in the USA where the scarcity of labour in industry had already caused the mechanization of its production on a scale unknown in Britain and Europe. Servant shortages were to encourage capital investment in the production of new machines which would make up for the absence of willing and able home labourers, and provide the housewife with a means of performing her own household tasks with ease.

Although, on one level, this 'substitution' argument holds true it is in fact a simplification of what actually occurred. Labour saving devices did, finally, fill the gap previously filled by servants, but the impact of these devices took some time in achieving widespread effects. It was not until a number of moves, made to improve the servant situation, had failed, that housewives began to utter such comments as the one made, in 1912, by the wife of a New England physician: *"I use a gas range, a fireless cooker, have an excellent vacuum cleaner, and an adequate supply of all kitchen utensils and conveniences. My household expenses have been cut down about 500 dollars a year, and I know of no easier way of saving that amount than by being free from the care and annoyance of a maid."*

The first reaction to the 'servant problem' was one of general dismay and attempts were made to improve conditions in the household in order to attract more home workers. One way of doing this was to make their work potentially easier by providing better, and more efficient, tools for servants to work with. This made housework more competitive with the other forms of employment – factory and shop work and work in commercial laundries where hours were shorter and physical labour was generally easier. Many labour-saving goods were bought, therefore, as a means of *keeping* rather than *replacing* servants. The fact that servants were a form of social status for the expanding numbers of middle-class

families at this time meant that they could not be disposed with that easily.

In spite of the use of appliances as an incentive for servants, there was still a great tendency to retain handwork when servants were employed. The high cost of the utensils meant that they were only introduced when absolutely necessary, and numerous accounts from the period describe the back-breaking tasks of cleaning stairs with brooms and ironing with heavy sad irons. As David Katzman concludes in his book *Seven Days a Week: Women and Domestic Service*

AEG fruit press showing a servant operating the machine in 1911.

in Industrializing America, "The presence of servants probably retarded modernization in the household, and domestics would benefit little from its belated appearance in the home."

Although the figures of servants between 1830 and 1930 in the USA indicate growing numbers (from 960,000 to 2,000,000), this must be set against the huge expansion in the demand for servants from the growing urban-based, middle-class population. The decrease is better indicated by the fact that in 1920, servants were available to half as many households as desired them as in 1870. The first sign of change was the tendency for the white, native-born servants to be replaced by black and immigrant workers. This was followed by the living-out servant becoming a much more common phenomenon than the traditional 'live-in' domestic helper; and finally, by the 1930s, the concept of the 'cleaning lady' had replaced the 'servant girl' in all but the most aristocratic of households.

The shortage of servants must be seen, at least in the USA, as a direct influence on the fact that that country was the first in developing, and making commercially available, a huge range of labour-saving devices aimed at the middle-class household. The first decade of the 20th century saw the slow acceptance of the new products. Electric washing-machines, irons and coffee-percolators were among the first appliances available on the market, but their relatively high prices prohibited widespread use for at least a decade. Sears Roebuck published its first catalogue of electrical goods in 1900, but it only had a limited appeal since, as well as the expense, much of rural America was not yet electrified. By 1913, however, the Western Electric Company was producing and selling electric coffee-percolators, toasters, irons, washing-machines, warming-pads and vacuum cleaners to an expanding clientele. The American consumption of electrical appliances was such that while in 1900 only 3,000 washing-machines were produced this number had risen to 500,000 in 1919 and to 882,000 in 1925. Even as late as 1929, though — at the end of the first real decade of the mass-consumption of electrical appliances — Henry Ford was to argue that, *"There is some machinery to use in the kitchen today. We have the vacuum cleaner, the various electrical appliances, the electric washing-machine, the electric ice-box: but most of it is still too expensive."*

Although the idea of the 'servant-less house' was the subject (and title) of the English author, Randal Phillips's book of 1920, the 'servant problem' in the first decades of this century, was by no means as extreme in Britain. Numerous explanations

The 'Electric Servant' washing machine, 1934. The square cabinet includes a food mixer and attachments for a radio and rotary ironer.

have been given for the fact that although they decreased in numbers, servants continued to exist in many middle-class homes until the 1950s. One reason is linked to the idea that there was a greater sense of respectability attached to the job in Britain than in the USA and another was the greater increase in the middle-class population of the latter country. In her article 'Daughters and Mothers — Maids and Mistresses: Domestic Service Between the Wars', Pam Taylor has explained that although

more servants lived out in this period, it was still seen as a major occupation for many working-class girls. She stressed the social rôle that such employment performed: *"The fact that uniform was insisted upon until 1939 suggests the importance of servants as symbols of status"*. Until labour-saving devices could fulfil this same social rôle there was to be widespread resistance to them.

Although writers such as Randal Phillips tirelessly reiterated this 'substitution' ideal, there was much evidence that, in many cases, labour-saving devices were purchased by the same households which could also afford servants. One was not necessarily seen as a replacement for the other.

An English book, written in 1926 by Ruth Binnie and Julia Boxall entitled *Housecraft: Principles and Practice* contained, for example, sections on the new appliances, in particular the vacuum cleaner, the gas and electric cooker, and the electric washing-machine. It also included a section entitled 'Servants and Employers' thereby suggesting that both appliances and servants co-existed in the same household. Binnie and Boxall encouraged the housewife *"to interest the servant in various kinds of handwork and give her an opportunity of reading daily papers and good books"*. Similar books from the 1930s continued to discuss labour-saving appliances and servants in the same breath. Aimed, in general, at middle-class and lower middle-class homes the assumption was that, in an ideal situation, they could co-exist happily. Even as late as the 1940s, vestiges of this attitude remained, although the 'daily help' was by now discussed more frequently than the live-in servant.

The British appliance industry had begun developing fairly quickly on the heels of its American counterpart and many of its products were indeed available by the first two decades of the century. As in the USA, however, the restrictions on consumption were a combination of price, access to the electricity supply and the rôle of these objects in middle-class life. Their widespread use, then, was delayed until after the Second World War by which time servants had greatly decreased in numbers.

In comparison with the USA, there was an enormous and longer-lasting resistance in Britain. The appliance companies were aware that until their products could achieve the dual function of fulfilling both practical and social needs, they would not replace the servant. As a result, they focused their advertising campaigns on the idea that possession of the new goods would bring with them enhanced social standing. By the 1950s, many people had been convinced by such propaganda and began to consume the new products as if there was no tomorrow.

The changing domestic rôle of the housewife is also closely associated with the increased use of appliances. The concept of the 'housewife' originates in what feminist writers have named the 'separation of the spheres' – the move within industrialization whereby a distinction was made between man's work, essentially outside the home, and woman's work, which remained domestic. This was a new development because women had played as vital a part as men in the production process when production had been in the home and not in the factory.

A number of feminist writers claim that the mechanization of the housewife's work has turned the rôle of the housewife into an alienated one. Also as Philip Bereano, Christine Bose and Erik Arnold have written in their article published in a book entitled *Smothered by Invention*, contrary to popular opinion: *"The time*

Brochure for Fisher autowash machines showing housewife in charge of operations, 1960s.

women spend on housework has not declined very significantly in the last 50 years despite the increased availability of modern 'conveniences'. Rather new household technologies have helped maintain the number of hours required for household tasks, causing technological unemployment among women domestics, seamstress and laundresses." The continued emphasis upon separate appliances for each household, instead of collectively-owned products, or communal services, has in addition, these writers suggest, reinforced the isolation of the housewife in contemporary society: "*Modern types of domestic technology tend to reinforce the home system, keeping women economically marginal to the larger society.*"

According to these and other writers on the subject, domestic appliances brought with them the need for ever higher standards of cleanliness, thus creating, as well as decreasing, housework and thereby keeping the housewife in her home. It was only with the expansion of women's paid employment outside the home that the amount of time spent on housework actually decreased.

The popular myth that household technology led to an increase in women's paid work needs some re-examination, however. As Ruth Cowan Schwartz has explained:

"*Housewives began to enter the labour market many years before modern household technologies were widely diffused; and the housewives then entering the workforce were precisely those who could not afford to take advantages of the amenities that existed.*" It is only in more recent years, with middle-class women increasingly seeking paid employment and appliances dropping in price, that we can begin to say that their 'labour-saving' qualities have had some influence upon this situation. Ironically, however, with the increasing social status associated

Brochure for English Electric washing machine, early 1960s. The image of the housewife is inspired by American examples.

with domestic appliances, more and more women go out to work simply as a means of being able to afford more consumer goods. They have become, in this instance, not an aid, but the end in sight.

The myth of the 'new woman', freed by domestic appliances was, essentially, created by advertisers in the inter-war period as a means of selling the new appliances. In his book *Only Yesterday*, of 1938, the American writer F. L. Allen was clearly taken in by it, when he wrote that:

"Much of what had once been house- work was now either moving out of the house entirely or being simplified by machinery...electric washing- machines and electric irons were coming to the aid of those who still did their washing at home...the housewife was learning to telephone her shopping orders, to get her clothes ready-made...to buy a vac- uum-cleaner and emulate the lovely carefree girls in the magazine adver- tisements who banished dirt with such delicate fingers. The 'new woman' had arrived personified by, and synonymous with the housewife and liberated from daily chores."

After the Second World War, when women increasingly became consumers as well as users, the adver- tising angle changed somewhat, and the glamorous 'hostess' housewife

emerged in these years, straight off the Hollywood film sets. Domestic appliances became desirable con- sumer durables and, as the English social historian, Harry Hopkins wrote in his book *The New Look*, *"It is the kitchen rather than the boudoir or the drawing room which is the heart of the feminine dream."*

There are clearly many fascinat- ing and complex issues to examine in the relationship between domestic appliances and the changing rôle of women in this century. Regrettably there are too many to go into in this short study, but suffice it to say that Giedion was right when he said that

the status of American women and domestic servants were linked to the mechanization of the home. It was not, however, a straightforward relationship, as he was implying, but rather a more subtle one in which the notion of 'cause and effect' is difficult to disentangle. Where the appliances are concerned, the most significant transformation is from their being seen as 'labour-saving' items to their becoming important status symbols – part of the concept of 'con- spicuous consumption' – a transfor- mation brought about by advertising and design.

A Kenwood brochure for a child's food-mixer from the 1950s which still describes the product as a 'servant substitute'.

2

KITCHEN PLANNING
AND APPLIANCE DESIGN

"Kitchens are as much a focus for industrialized work as factories and coal mines are, and washing machines and microwave ovens are as much a product of industrialization as are automobiles and pocket calculators."

Ruth Cowan Schwartz

ONE of the major influences upon the changing nature and forms of this century's appliances – particularly those involved with the preservation, preparation and cooking of food – has been their evolving relationship with the modern kitchen. This in turn has been largely influenced by the changing status of domestic servants and the housewife.

The idealized image of the domestic kitchen has undergone several dramatic character changes in the last 100 years. It has altered from a large workspace-cum-living area, made up of separate elements and centred around the fixed range, to a small laboratory-inspired room consisting of unified surfaces, and back again to a living space and focal point for the household. In more recent years, the increasing number of powered appliances crowding the floor area and working surfaces has meant a radical revision of the ways in which kitchen furniture and equipment relate to each other.

The subject of kitchen design has preoccupied many people from a number of quite distinct areas since the middle years of the last century, among them architects, feminists and 'time and motion' experts. In 1841, for instance, Catherine E. Beecher, the Principal of the Hartford Female Seminary, published her 'Treatise on Domestic Economy for the use of Young Ladies at Home and at School'. She outlined a number of ideas for organizing work permitting women the greatest amount of freedom and providing maximum efficiency in the home. She advocated the need for a 'light, neat and agreeable' kitchen, although in her plan for a 'good kitchen' she retained the traditional separation of the scullery from the stove-room. She was unable, of course, at that early date, to recommend the use of electrical labour-saving tools.

Even 70 years later Christine Frederick, in her book *Scientific Management in the Home*, first published in instalments in the *Ladies Home Journal*, listed a range of kitchen tools – among them an egg-beater, a mixing bowl, a nutmeg grater, a can-opener, and an apple-corer – which are all of the hand-powered variety. She did, however, make references to the advantages of both gas and electric ovens, in particular *"the latest models which have ovens either at the side or above the*

The Model Electric Kitchen from the Chicago Columbian Exhibition of 1893.

ciently around it and proposed the standardization of the heights of the working surfaces. Her main preoccupation was with 'step-saving' or 'routing', a term borrowed from contemporary scientific management studies on work in factories which proposed the idea of breaking down work into single tasks. (This mitigated against the use of complex labour-saving appliances in the home such as the electric food-mixer which performed a number of tasks.)

In a preface to the text, Frank Gilbreth, an important advocate of 'scientific management', wrote that *'Mrs Frederick has seen the necessity for making the home a laboratory'*. This observation became highly influential in subsequent years. Efficiency could be measured scientifically according to this method which appealed to many people seeking ways of rationalizing work at home. What Mrs Frederick did not take into account were the psychological problems inherent in isolating the housewife in this new, separate, efficient workspace. The consequences were not to be fully felt, however, for another two decades.

The implications of the 'laboratory' kitchen for electrical appliances were as yet fairly minimal since the kitchens depicted in Mrs Frederick's book still consisted of isolated elements, whether work table, utensils, or shelving systems. They were

regular cooking surfaces". Mrs Frederick's work was the proof of the idea — evident also in the story of the rationalization of modern industry — that re-organization precedes mechanization. In this context she stressed the desirability of several changes — for instance, the move to the smaller, self-contained kitchen, used solely for the preparation of food and opening directly on to the

dining area. Also she suggested the sequential organization of equipment and furniture so that the refrigerator, preparation surface, cooking surface, and serving surface were positioned in a line. In addition she proposed the abolition of the central table and the inclusion of a tall, adjustable working stool. She used the analogy of the 'mechanic's bench' with its tools grouped effi-

simply brought together into a smaller space and grouped logically. The problem of the fixed range was solved by using it as the starting point of the arrangement and more thought was given to the materials used on the walls and floors than to the nature of the tools used, which remained fairly traditional. Laundering was simply relegated to another area of the house. One innovation with implications for appliances was the inclusion of the refrigerator in the kitchen. Hitherto it had been relegated to the verandah, both because of the smell of ammonia which was used as an early refrigerant and for accessibility for the ice-man who delivered ice on a

An image from the 1930s of an efficient British kitchen modelled on factory lines.

weekly basis. The 'step-saving' principle put priority on the position of the utensils, whatever the disadvantages of the arrangement.

A spate of books on the, by then, fashionable subject of Household Management appeared in the USA around the First World War, most of them based on Mrs Frederick's writings and sharing her ideals. In 1914 Mary Quinn, in a book entitled *Planning and Furnishing the Home* supported most of Mrs Frederick's ideas but felt, nonetheless, that the refrigerator was still best placed in the vestibule and that, although cleaner to use, the electric range was out of the financial reach of the majority of households. She felt that Beecher and Mrs Frederick were clearly aiming their texts at the middle-class household, making few allowances for lower incomes.

From the basis of the efficiently organized kitchen with its open shelving and visible displays of hand tools grouped on the wall, grew the more sophisticated 'fitted' kitchen, instigating a system of continuity in its work surfaces and overhead storage. Inspired by the ideas in Mrs Frederick's book, many of the European Modern Movement architects of the 1920s developed her concept of rationalization into plans for kitchens based on a logical work process and a system of standardized fitments.

Among the numerous European attempts to emulate the American pioneering work in this field were the designs of Ernst May and Grete Schütte-Lihotzky's team of architects and designers in Frankfurt. These stood out as the most thoroughgoing of them all and the 'compact' kitchen that they evolved for the rehousing plans for that city were based upon Mrs Frederick's basic principles. They took the principle of standardization even further by applying it not only to the height of the counters, but to that of the cooker also. Their emphasis on closed, unified storage units rejected Mrs Frederick's concept of exposed, grouped tools. The tendency now was to see the kitchen as a set of flush surfaces with equipment and food neatly stored behind closed doors.

One of the problems that this raised was that of storing the new small, powered appliances beginning to appear in Germany on quite a wide scale. While traditional kitchen tools – knives, ladles, pots and pans, etc. — were relatively easily stored, it was less easy to conceal a bulky electric toaster, coffee-percolator or floor-polisher in the small cupboards provided.

Thus a fundamental question was raised which has been frequently reiterated since — how can the principle of standardization be applied to small electrical appliances? On the

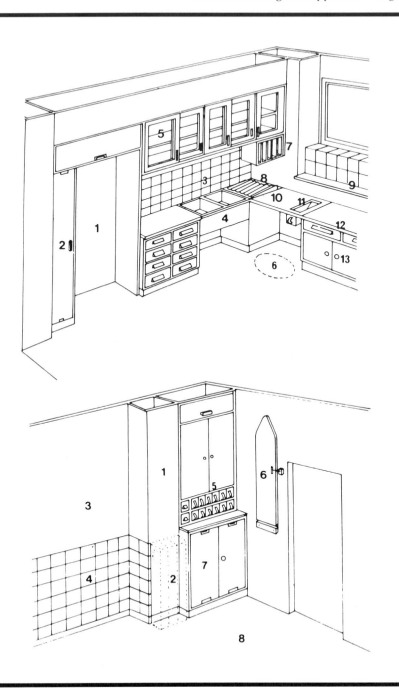

whole, small appliances have never fitted into purpose-built cupboards but stand, instead, on the surfaces of kitchen counters. This lack of unity between kitchen furniture and small appliances is emphasized by the rôle of appliances as status symbols. To truly represent 'conspicuous consumption' they *must* be seen. From the 1920s to the 1960s, therefore, small kitchen appliances incorporated an aesthetic which stressed 'display' rather than efficiency or rationalization.

Once the mechanization of the kitchen joined its re-organization, the issues involved in creating the modern kitchen became increasingly complex. The tension that emerged between the structure of the kitchen and the appliances within it, determined many of the changes that occurred in the evolution of the former since the 1920s.

By the 1930s, the small 'laboratory' kitchen advocated by Mrs Frederick had influenced a range of different architectural designs — from the new suburban housing developed in Great Britain, to the show houses at Chicago's Century of Progress Exhibition of 1933. Here

The efficient kitchen, modelled on Christine Frederick's example, designed by Ernst May and Grete Schütte-Lihotsky for the Frankfurt housing scheme in 1927.

the kitchens were described as 'science laboratories which could be washed down with a hose'. Where major appliances were concerned, this type of kitchen emphasized small, standardized units, most of them now with white, or off-white, enamelled surfaces conforming to the ideal of cleanliness and efficiency which these kitchens stood for. The concept of 'white goods' grew, there-

fore, out of the 'laboratory' kitchen and belongs to the period between the two World Wars. The minor appliance manufacturers continued, however, to fight against conformity and the goods remained as emphatic as possible. In a book of 1935 entitled *What's New in Home Decorating?* the American writer Winifred Fales explained that, "*well-known artist-designers have modelled proud-look-

The kitchen in the Electrical Association for Women's 'All Electric House' in Bristol, 1930s.

ing percolators, aristocratic kettles and casseroles with definitely modern contours."

There were some major appliances which remained physically independent through their exaggerated scale and swollen forms. For instance, by the 1930s, the refrigerator in the USA had developed into a bulbous monster having more in common with the automobile parked outside than with the modular kitchen units that it stood alongside.

Despite the American influence, European manufaturers acknowledged the need for compact, fitted appliances. This was most apparent in Germany where the system of standards established by the Deutsches Institut für Normung between the Wars, instituted a compatibility system between kitchen furniture and equipment which has run through this century and set a precedent emulated by many other countries.

While not fully applying strict principles of standardization, Britain also appreciated the need for compactness and pioneered the concept in a number of its appliances. Belling's 'Modernette' oven of 1919, for instance, was intended for the

The 'Admiral' electric refrigerator, 1950.

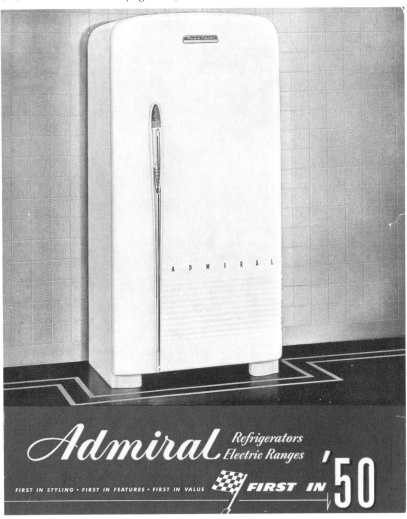

The 1935 Frigidaire. Its bulky forms prevented it becoming an integrated element in the fitted kitchen.

most compact of kitchens, and in 1946 British Electrolux developed a small, fitted fridge to go into the Government's post-war, temporary housing. Even in the USA, an attempt was made by the Acme-National Company in 1950, to emulate the Electrolux example. It produced a refrigerator in the form of a set of drawers, only 27 inches wide and 36 inches high. A writer in *Design* magazine commented that it was intended for use by the "*struggling housewife whose whole living quarters could be slipped into one Hollywood kitchen, with room to spare.*"

Just as the USA took the kitchen to one extreme, with the 'laboratory' model that country soon afterwards, pioneered the 'live-in' kitchen. After the Second World War it became an increasingly popular alternative to the 'laboratory' kitchen. By the mid-1930s, with the final demise of the live-in servant in many homes, most American housewives did their own housework. There was an increasing need therefore, for the kitchen to become more linked to the rest of the house and to become a pleasant environment as well as an efficient workspace. In the smaller urban apartments, the kitchenette replaced the small kitchen and dining-room and, in many new houses the dining-room was increasingly eliminated and combined with the kitchen. The reasons

for this were twofold — while it was a direct result of the move towards smaller houses and the need to combine spaces, it was also derived from the housewife's desire to be part of the household while she prepared meals. As another means of 'humanizing' the kitchen, colour began to replace the neutrality of Mrs Frederick's 'laboratory', and patterns, otherwise reserved for living areas, began to take over from the dingy linoleum and tiles.

In 1935 Winifred Fales explained that:

"*Now that the kitchen bids fair to become the social center of the home, with the game of Raiding the Ice-box threatening to supplant Bridge as a popular diversion, our old nondescript collections of pots and pans, unrelated pieces of apparatus and the cream walls and white scrim curtains are emphatically back numbers... Today's kitchens are excitingly dramatic.*"

She went on to describe how the rôle of Science in the kitchen had been joined with that of Art and that colour and pattern had combined forces with the most advanced technologies and materials. In her new kitchen interior she included a book stand, a radio and an orange coffee-percolator to add a dash of 'brilliance'. Importantly, a breakfast

Belling's compact electric 'Modernette' cooker, 1919.

table was part of this new 'humanized environment'.

The unification of the 'working' with the 'living' space had all sorts of implications for kitchen appliance design. In 1945, in their book *Tomorrow's House*, for instance, the Americans Nelson and Wright recommended a 'service opening' between the kitchen and dining room. This had the effect of putting the percolator and toaster "*within reach of the table, giving the master of the house something to do while he is waiting for his egg.*" The idea of the breakfast bar also became a very popular means of dividing the cook-

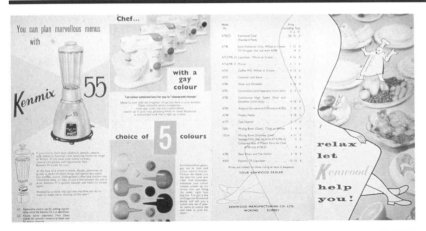

Brochure for Kenwood's 1955 food-mixer showing a range of coloured plastic attachments in yellow, red, blue, green and black.

ing from the eating and relaxing areas. The main implications for appliances was that, as highly visible objects, they increasingly turned into fashion items rather than the useful tools they had largely been until then. It was not to be long before they incorporated colour and flamboyance into their appearance.

The way in which ideas about kitchen planning affected various levels of society in different ways at different times was well documented at the 'Britain Can Make It' exhibition held in London in 1946. Four kitchens were shown as a means of highlighting the options available to different strata of society. They included F. Gibberd's 'flat kitchenette' equipped with a fitted oven, an Electrolux fitted fridge and a counter separating the eating area; a luxury, self-contained kitchen designed by Maxwell Fry and Jane Drew, containing a separate hob and eye-level oven and had a number of

A kitchenette in a small flat, designed by Frederick Gibberd for the 'Britain Can Make It' exhibition of 1946. It contains Electrolux's fitted fridge.

appliances on its extensive surfaces; a middle-class kitchen with a dining recess, designed by F. MacManus; and a more traditional living-room kitchen, designed by Mrs Braddell, which had a built-in solid-fuel range with a Windsor chair beside it.

It was becoming apparent, by the 1950s, particularly in the USA but also in Britain and Europe, that because of the number of appliances in homes new ways of thinking about the kitchen needed to be formulated. A debate ensued focusing on the old question of whether or not appliances could be standardized or whether they were still essentially independent, status objects. In Britain, the architects Alison and Peter Smithson suggested the idea of the 'appliance house' which would be both 'technological and cosy'. The main problem they isolated was that of the clutter presented by appliances in otherwise ordered kitchens, and the Smithsons proposed a 'concentration of appliances' which were mobile and could therefore be brought into use when needed and put out of sight when not. The architect Eric Lyons joined in the discussion that ensued claiming that the 'attempt at standardization' had been quite useless. *"It has mislead everybody, and it is much better to come clean about this and regard appliances as self-contained little pieces of movable furniture."*

The debate about 'fitted' or 'independent' appliances dominated most of the discussions about the relationship between kitchens and their appliances after the War. In the USA, the 1950's 'dream' kitchen, with its automative-inspired, free-standing appliances which, according to the designer Henry Dreyfuss, 'give you the feeling that you could climb in and drive away' were gradually replaced by the new 'packaged' kitchens. Their 'fold-up', 'tuck-in' units succeeded in disguising the oven and other pieces of electrical equipment. The 'built-in' look for refrigerators meant the end to bulbous curves and an emphasis, instead, on straight lines. Removable front panels meant that appliances could be matched to the kitchen's decorative scheme.

Another American innovation of this period was the free-standing unit which, looking like a room divider, was made up of cooking, serving and freezing sections. One such unit, a six-foot 'Gourmet' serving centre, was described as being *'finished in walnut and antique leather, to emphasize its living-room appearance'*. These were early moves in the direction of making pieces of kitchen equipment look part of the living space of the house, a tendency which re-emerged as one of the available options in the 1980s once 'built-in' had become a standard feature of most major appliances.

By the 1960s, a greater sense of

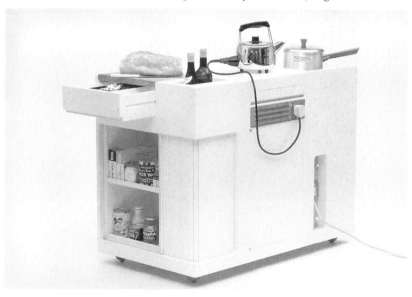

A mobile kitchen capsule, 1965.

visual unity had emerged in major appliances: cookers, fridges and washing-machines increasingly conformed to the basic principle of the 'white cube'. It proved, however, in Britain at least, difficult to persuade manufacturers of the need for standardization and much size variation remained. This became a burning problem again in the 1960s and all sorts of suggestions were put forward to counter it. One such proposal was Allied Ironfounders' 'Service Wall' which consisted of a central core housing all the electrical services. Everything was controlled by a central panel. Another British experiment from the same decade was the 'Capsule' kitchen which was attached to a central revolving cylinder housing all the essential functions.

Such futuristic proposals aside, the general international tendency from the mid-1960s onwards has been for built-in kitchens with appliances integrated as much as possible. Only a minimum of small appliances – a food mixer and a kettle, say, – are left on the continuous surface. These came in the 1980s, in rustic, 'natural' colours or in traditional white. A basic contradiction remained, however, as the move towards increased numbers of small, differentiated appliances in the kitchen, combined with the growing idea of the kitchen as a social centre, argued against the unified, flush kitchen with no messy protruberances. As Mary Alexander argued in *Design* magazine in 1984, "*kitchen planning is still proudly described as a 'matter of logic' whereas the primary need is for it to be responsive to changing lifestyles and changing patterns of use.*" The discussion continues although kitchen furniture makers and appliance manufacturers still tend to work in isolation from each other.

An electrically controlled circular kitchen designed by George Fejer.

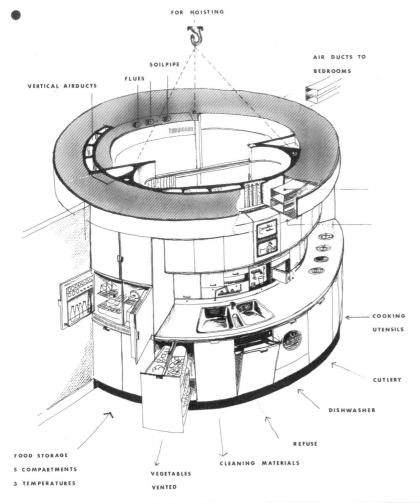

3

THE ELECTRICAL APPLIANCE INDUSTRY

IF the demand for domestic appliances has been determined by a number of social factors, their supply is equally constrained by technological and commercial pressures.

It is difficult to provide a precise date for the emergence of an industry dedicated to the production of electrical appliances for the home. It emerged, like so many other modern industries, from several divergent routes. In the USA, some of the first companies were formed in the 1880s, 1890s and 1910s with backgrounds in two main areas. These were the new electrical engineering industry (the commercial application of the discoveries made by Thomas Edison) and a group of smaller, 'light' engineering manufacturers which turned from other products (in particular ones fabricated in metal) to concentrate on the manufacture of electrical appliances. Into this latter category fell such well-known American companies as the Sunbeam Corporation – which began by making scissors and clippers for shearing sheep; Landers, Frary and Clark (better known by the 'Universal' brand name) – which moved over from cutlery-making; Maytag, which made farm implements; and Hoover, which had hitherto been involved in making leather goods. All of them turned to the manufacture of what they considered to be more profitable products. A little later, two automobile companies – General Motors and American Motors – also diversified into appliances at a time when they felt they needed a new consumer product to lift them out of recession. The pattern of these types of companies moving into electrical appliances has been echoed in countless other countries through this century.

The electrical supply industries have also had a vested interest in the manufacture of electrical goods – another important factor in the production of appliances. Ruth Cowan Schwartz has pointed out that many of this century's most popular domestic appliances are ones which it has suited the electrical engineering industry either to produce or sell.

Adrian Forty, in his article 'Electrical Appliances 1900–1960' published in a volume edited by Thomas Faulkner and entitled *Design 1900–1960* has supported the idea that the alliance between the supply industries and the appliance industries has been an extremely strategic one: "*Electrical engineers found that to reduce the costs of production and therefore the price of electricity it was essential to increase the ranges of uses to which electricity could be put and also to increase the numbers of consumers.*"

One general characteristic that has underpinned the success of particular companies in the appliance sector has been their ability to invest vast sums of money into the establishment of a mass-production system. As a result, the giant companies which have dominated the electrical industry have had a strong profile in the appliance field as well, gradually absorbing the smaller light engineering companies. Companies such as General Electric and Westinghouse in the USA; AEG and Siemens in Germany, and The General Electric Company and Thorn EMI in Britain, have succeeded either in taking over, or putting out of business, dozens of their more modest competitors which could not afford to keep up with their level of research, their scale of production and the sophistication of their marketing techniques. Only those smaller manufacturing companies have survived which had sufficient capital behind them from another product or products. Many others only lasted through the early 'workshop' days of the industry and fell by the wayside later. Since the 1920s, the electrical appliance industry has been large scale and capital intensive. It has had to develop aggressive marketing techniques as a means of persuading the ever-expanding mass-market both of its need for their products and of their superiority to those of their competitors.

The USA and Germany led the way in the early decades of appliance industrialization while a number of other European countries – notably Sweden, Holland, France and Italy, and eventually Great Britain – followed their lead. The period after 1950 is dominated by the resurgence of Germany and the spectacular rise of Japan as a force to be reckoned with.

Because of its early and continued commitment to mass-production and to standardized products the appliance industry has always sought world markets. As a result, the large American companies were quick to establish overseas divisions. The ubiquity of the 'Hoover' vacuum cleaner, 'Electrolux' refrigerator, 'Braun' food-mixer and, more recently the 'Sharp' microwave is a direct result of the continued international policy of aggressive marketing followed by the electrical manufacturers. From the 1880s onwards, the availability of electrical goods on the American market was due, in part, to the efforts of the expanding electrical industry to supply products to use up their electrical power. They were

An advertisement for a range of Universal's small appliances from Ladies Home Journal, 1909.

mainly preoccupied with early forms of electric lighting and, in the 1890s, with the application of electricity to tramways and railways. Gradually, however, they began to diversify their interests and to manufacture electrical appliances. First, in the 1880s, came electric fans, run from overhead light sockets. They were followed, at the turn of the century, by a number of small appliances, such as irons and toasters, as well as small electric motors which could be connected to sewing– and washing-machines.

The American electrical engineering industry had not set out with these ends in view, however. The largest company, General Electric, was formed in 1892 from a merger of the Thomson-Houston Electric Company and the Edison General Electric Company, both of which had concentrated on lighting and its necessary equipment and on heavy electrical machinery. The other American giant, the Westinghouse Electric Company, formed in 1886, shared the same preoccupations through the 1880s although it had originally worked on air-brakes for the railroad.

Based on inventions and scientific advances, the American electrical industry was established by a number of 'engineer-entrepreneurs' who could see the commercial possibilities of the new developments. Aided by the growth of the American market in the 1870s and 1880s and the increased availability of unskilled workers and growing urbanization the industry expanded quickly. The companies' early financial successes lay in supplying electrical equipment – such as lighting fixtures, lamps and sockets – to both commercial and domestic markets. They also supplied large-scale apparatus, such as turbines and generators, to industry, the railroads and the public utilities. However, with the exception of fans, the first ranges of appliances were geared to industry and commerce rather than directly at the householder. Irons, for instance, were intended for tailors and commercial laundries and toasters for hotels, boarding houses and mass-catering facilities.

Although General Electric had produced a range of irons and toasters as well as an electric range by 1913, it wasn't until the 1920s that the consumer market made up a significant part of that company's sales. Between 1925 and 1930 it devised and put into production a whole range of new, 'labour-saving' electrical appliances – among them refrigerators, washing-machines, vacuum cleaners and numerous small appliances – designed to help banish drudgery in the home. As J. W. Hammond has explained in his book *Men and Volts: The Story of General Electric*:

"soon products for the home began to be almost as large a part of GE's business as the large apparatus...GE assisted in creating a whole new branch of the electrical industry, providing new employment for thousands of people in the factory, thousands of salesmen all over the country and a host of workers in other industries providing the raw materials from which these appliances were made."

This new venture coincided with the expansion and economic success of the American electrical appliance industry as a whole. By 1920, the public demand for these goods was still largely untapped but between 1921 and 1929 the output of the industry tripled in value. The 1920s was also the decade in which the large companies took over the lion's share of the profits, such that by 1929 there were only 11 companies in control of 85 per cent of the industry. The recession which followed increased this tendency, forcing many of the small companies out of business and confirming the power and status of the larger firms.

Although it was not until the 1920s that the large electrical companies turned their attentions to the world of domestic consumer appliances there had been many smaller, more specialist firms which made electrical appliances and marketed them directly to the householder prior to that. They produced relatively small quantities of goods, manufactured largely by hand by sheet-metal artisans. Selling was done by trying to persuade the power companies to take their products as electricity load-builders and by door-to-door salesmen who approached the householders in all the newly electrified towns, offering free trials and payments by instalments.

This latter form of selling became fairly common practice between the years 1900 and 1914, and the number of salesmen travelling around the country loaded with domestic electrical appliances increased enormously. The Hoover Company was the first to do it as a means of distributing its early vacuum-cleaners. Hoover's product was invented in 1907 by Murray Spangler who was bought up by the Hoover family. Originally in the leather and saddlery business, Hoover was looking for a new product now that cars were taking over from horses. By the following year the company had expanded enough to be able to afford a full-page advertisement in the *Saturday Evening Post* encouraging the domestic market to purchase its new product. The original 'Model O' Hoover was made very simply from sheet metal. Arthur Pulos has explained in his book *The American Design Ethic* that it was, "*essentially a 'shop-form'*

The 'American Beauty' iron, 1912.

product, constructed primarily by hand methods in a sheet-metal shop, with a brush roll and fan made of die-cast aluminium and an oil-treated sateen bag."

The Hoover Company's main contribution to the history of domestic appliances lay in its selling methods. Not only did the company salesman travel around from door-to-door, he was also trained to answer very detailed questions about the products and to give demonstrations. By the 1920s, the 'Hoover' man on his bicycle had become a familiar sight in the American suburban and rural landscape.

The turn of the century saw the formation of other well known, and less well known, electrical appliance manufacturers in the areas of small appliances, ranges and washing-machines — among them Hotpoint, Sunbeam, Maytag, and Landers, Frary and Clark. But it wasn't until

the First World War that the electric refrigerator was finally manufactured for the domestic market. This was achieved partly by the efforts of the General Motors Company to move into this major appliance area at the end of the War.

In 1918, the company's president purchased the Guardian Refridgerator Company of Detroit and within ten years he had turned an unsuccessful product into a market leader. While the original fridge had been made by one man with a bare minimum of tools the Frigidaire division of General Motors invested vast funds in improving the product. It changed its wooden cabinet into a steel one; lowered the price; and manufactured it in much greater quantities. In the 1920s the manufacturers were learning about steel technology and assembly line production from the automobile industry. As a result they were able to expand their production enormously. The fact that both the car and the refrigerator were basically boxes containing a motor reinforced the links between the two products and their means of manufacture. In the mid-1930s, Frigidaire added electric ranges to its production and following the Second World War diversified into a number of other related products including dish-washers, dryers, washing-machines, and ice-cube makers.

Sweep With Electricity For 3c a Week

You Can Afford *This* Electric Suction Sweeper As Easily As You Can Afford a Sewing Machine

No more dirt or dust! No more back-aches on cleaning day! This wonderful little machine takes up all the dust, scraps and dirt from carpets, furniture, curtains and portieres more perfectly than any of the big vacuum cleaners for the services of which you pay $35 to $50.

It works like magic. Simply attach the wire to an electric light socket, turn on the current and run the machine over the carpet as you would an ordinary carpet sweeper. Its rapidly revolving brushes loosen the dirt, and the strong suction pulls it into the dirt bag in the twinkling of an eye. Nothing escapes its marvelous cleaning influence.

So simple a child can do it.

So economical anyone can afford it.

Have your cleaning finished in one-fourth the time and with one-tenth the labor!

HOOVER Electric SUCTION SWEEPER

For All Houses Wired for Electricity. Price $70; Extra Attachments, $15 per Set

Makes Carpets Bright as Well as Clean

Unlike All Other Vacuum Cleaners

No Dusting Afterward

Dust is Full of Disease

Extra Attachments

Blows Up Pillows and Mattresses

Cleans House for 3c a Week

We Can Supply You Now

Free Trial

An Opportunity for a Few Dealers

FREE TRIAL OFFER
Hoover Electric Sweeper Co., Dept. 14, New Berlin, Ohio

Name

Street

City   State

Hoover Electric Sweeper Co., Department 14, New Berlin, Ohio

IN ANSWERING THIS ADVERTISEMENT PLEASE MENTION COLLIER'S

A whole page advertisement for Hoover's 'Model O' in the Saturday Evening Post, 1909.

The story of the rise of the electric refrigerator in the USA reflects the domestic appliance industry's move into mass-production. General Electric (GE) and Kelvinator (part of American Motors) were the other two giants who were ready, by the 1920s, to invest enormous sums of money into a product that was destined to enter every home now that electrification was more widespread. In *More Work for Mother* Ruth Cowan tells in detail the story of General Electric's investment in the electric compression refrigerator. She points out that by 1923, although there were a total of 56 companies in the refrigerator business, "*only eight were either well-financed or well on the way to large-scale production*". The key to the success of GE's 'Monitor Top' fridge lay in the gigantic advertising campaign that accompanied it, launched by the GE department created especially for that job. The dominance of the electric over the gas fridge was mainly due to GE's efforts to promote it.

Another significant aspect of the growth of the American appliance industry was the rôle played by the mail-order houses, Sears Roebuck and Montgomery Ward. By the 1930s both had become important manufacturing forces in this area.

The American appliance industry, as we know it today, was formed in the boom years of the 1920s and

Frigidaire's refrigerator, 1927.

quick to understand the economic advantages of supplying the equipment necessary for electrification. Siemens & Halske was formed in 1847 and through its involvement with the development of the telegraph system, dynamo, telephone and lighting, became a powerful concern by the end of the century with subsidiaries in both the USA and Britain. Emil Rathenau formed Deutsche Edison Gesellschaft after he had seen Edison's inventions in Paris in 1881. Four years later, under the new name of Allgemeine Electricität's Gesellschaft (AEG) he turned to manufacturing. By the early 1880s Siemens and AEG were in direct competition with each other. Together they dominated the German

General Electric's 'Monitor Top' refrigerator, 1934.

shaken into place by the recession of the following decade. The 1950s saw a second period of expansion during which time the giants from the inter-war years continued to go from strength to strength. General Electric, Westinghouse, Frigidaire, Sunbeam, Maytag, Hoover, Kelvinator and Bendix (a late-comer on the scene which came out of aircraft manufacturing) became household names after 1945. The model presented by these companies influenced the experiences of those in other countries as well, albeit with a few variations.

The German electricity industry, for instance, developed in the second half of the 19th century through the efforts of two large-scale companies

Kelvinator's 'Model 299' refrigerator, 1922.

electrical engineering industry. Like GE and Westinghouse, these giant concerns were relatively slow to enter the field of domestic appliances – as in the USA, this was left to a number of smaller metal manufacturing firms. When they finally entered the market, Siemens and AEG proved to be forces to reckon with.

Siemens had manufactured airships, airplanes and cars before it finally established a division in 1920 named Siemens Electrowärme GmbH (SEG) concentrating on the production of electrical heating equipment and home appliances. The rationale was identical to that in the USA although it was not until this date that a market had emerged that was large enough to warrant the heavy investment needed for their mass-production. In the 1920s, Siemens adopted the trade mark 'Proto' (used originally in 1908 for its cars) for electric cookers, vacuum cleaners, washing-machines and irons manufactured by SEG and put on the market by the end of the decade. The products were highly utilitarian in their appearance, representing an aesthetic which soon came to characterize German domestic appliances.

This division of Siemens was dissolved in 1945 but it re-opened in 1957 under the name Siemens Electrogeräte ready to meet the needs of the German post-war consumer boom. It added radios and televisions to its list of items of domestic equipment and today the German company Bosch has combined with it in its manufacture of electrical appliances for the home. The company, however, never saw domestic appliances as anything other than a

small area of its concerns, and it remains one of the largest electrical engineering companies in Germany today.

The rôle of the domestic appliance plays a somewhat higher profile in the preoccupations of AEG since it began to manufacture them at an earlier stage in its evolution con-

Montgomery Ward vacuum cleaner designed by W.D. Teague, 1939.

The 'Cosyglo' electric heater manufactured by Siemens in 1926.

sidering them as highly marketable commodities. AEG's reputation in this area is largely due to its pioneering step in bringing in the well-known German designer, Peter Behrens, in 1907 to work on arc-lamps, and then on other products such as fans, kettles, and coffee-pots. The Behrens products were not, however, the company's only foray into electrical appliances. At the turn of the century AEG already had more than 80 appliances on offer. In 1889, for instance, an electric cigar-lighter and tea-kettle had been presented at the Berlin exhibition of that year; in 1894 a coffee-machine was added to the list, available in either 'Renais-

sance' or 'Rococo' style; and in 1899 an electric hair-dryer was introduced as well. The brand name of 'Dandy' was applied to a range of goods, including a vacuum cleaner in 1911, the year in which an electric fruit-press was also introduced; and in 1912 an ice-machine was included in the list of household goods. In the early 1920s, AEG's 'Vampyr' vacuum cleaner was launched on the market, an aggressive-looking machine which deserved its formidable name. By the 1950s, a whole range of objects was on offer, including fridges, and automatic washing-machines.

AEG helped Germany acquire a prominent position in the evolution and design of domestic appliances. Today Siemens and AEG still control the industry and have financial and production links with many companies associated with high design standards in the domestic appliance field, among them Bosch and Neff.

Another company which also helped Germany acquire its reputation after the Second World War was Braun. Established in 1951, it promoted very high standards of design and finish in its products, extending from audio equipment to appliances such as fruit juicers and food mixers. In the mid-1950s, Braun employed a number of free-lance and in-house designers, to ensure the high quality in its products and the aesthetic that the company encouraged quickly

became admired and emulated on a world-wide basis. Thanks to the efforts of Braun, in the late 1950s and 1960s, Germany became the leading country for high standards in domestic appliance design.

During the last decades of the 19th century, the British electrical industry was dominated by the large American and German companies which had opened subsidiaries on British soil. Despite the number of technological advances in this area made by British citizens (among them Michael Faraday, who had discovered the dynamo, and Swan, who had worked on the carbon filament lamp), GE, Westinghouse and Siemens were the powers behind the early years of electrification in Britain. It was they who dominated the process of electrifying both the tramway system and the London Underground. British industry was slow to exchange its steam-powered machinery electricity and, generally speaking, the rates of urban and income growth were slower in Britain at this time than in either the USA or Germany.

The emergence of an electrical appliance industry was also slower and on a smaller scale than in these countries. There were many reasons for this, among them the importance of gas as an alternative source of power; the delayed standardization and expansion of domestic electrifi-

An absorption refrigerator manufactured by AEG in 1926.

cation; and the relatively slow expansion of the demand for labour-saving devices in the home. This did not, however, prevent the emergence of a number of small-scale manufacturing companies interested in domestic electrical appliances. These were, as elsewhere, generally small-scale, light engineering firms with a background in metal technology.

The British General Electric Company began in 1886 from crude beginnings, hand-making to order such items as electric gas lighters, batteries, electric wiring and 'crude electric bells and gadgets'. But by 1900 it had moved away from the consumer market to become a large-scale electrical engineering company, only to return to produce domestic electrical appliances after the Second World War. Another company established in the 19th century – Crompton Limited – began making arc lamps, generators and electrical equipment on a modest scale in 1878. Formerly the chief engineer in Stanton iron works, Crompton was highly conscious of the need for products to act as load-bearers for the electricity supply during the day. Together with E. J. Fox and H. Dowsing, he went in to the small-scale manufacture of heating and cooking appliances aimed at both a commercial and a domestic market in the 1890s. It was from the Crompton works that, in 1912, Belling emerged to set up his own tiny concern in a shed in Enfield to manufacture an electric fire. He employed a fitter and a boy who according to one historian of the company *"commenced making benches and stools, fitting up a few vices, drills and other small assembling appliances"*. The early success of the company, later to become Belling & Co. Ltd, lay in the development of a fairly aggressive marketing policy involving the owner of the company travelling around the country on his bike armed with a catalogue and a sample, selling his product to electricity suppliers.

Such early initiatives were isolated in Britain, however. It was not until the inter-war years that the appliance industry began to expand

A circular hot-plate and spare top manufactured by Crompton & Company in 1891.

An electric oven manufactured by GEC in 1895.

appreciably , and not really until the 1950s that one could begin to talk about mass-production on anything like the American scale of operations. The Electrolux factory in Luton, opened as a British subsidiary of the Swedish company, was an early example of an attempt to introduce mass-production into the British domestic appliance industry. The company took over a World War I airplane factory and began to manufacture — first vacuum cleaners of the cylinder type, and then, by 1927, refrigerators aimed at the domestic market. Through the 1930s, Electrolux survived by selling its fridges to the gas industry and by employing the same 'hard sell' techniques with

its vacuum cleaners that Hoover had introduced in the USA at the beginning of the century. After turning to the manufacture of munitions during the Second World War (like so many other companies in this field), Electrolux expanded considerably in the second half of the 1940s as a result of a government contract to supply 50,000 temporary houses with built-in fridges.

During the 1930s, many of the domestic appliances available in Britain were foreign imports and a number of companies, such as Hoover and Hotpoint, even began making their products on British soil. A number of other companies also emerged from electrical engineering, and other backgrounds, becoming increasingly important after 1945. Among them were the EMI company, formed in 1931 as a merger between HMV and Columbia, which continued to operate a division called HMV Household Appliances Ltd, the English Electric Company, formed in 1919 but which didn't enter the appliance field until 1927; and Morphy Richards, formed in 1936 and which was quick to utilize a mass-production system. The industry developed largely as a response to foreign competition. While American companies manufactured the larger, more expensive products, such as fridges and washing-machines, the British companies moved predominantly into the areas of irons, vacuum cleaners and electric fires. Following the Second World War British Production capacity had increased and many firms moved from batch– to mass-production in response to the huge expansion in consumer demand. As a result, the industry moved into the hands of a few, large-scale concerns. Companies emerging included Thorn — later to become Thorn EMI. It systematically went about buying up many smaller companies, including Tricity, Moffat and, later, Kenwood — a company formed in 1947 which rose to a position of strength first through the production of toasters and then an electric food mixer which quickly became a market leader. Engineering companies, including Ada and Lec Refrigeration Limited, were also attracted into this area in those heady post-war years and by the late 1950s it was possible to speak of a British domestic appliance industry. As in the USA, the tendency was towards the domination of a few companies and by 1966 Hoover, EMI, Hotpoint, Belling and GEC accounted for 38 per cent of the total industry.

Miniature electric ovens moving along the production line at Belling & Company, 1960s.

Many companies had also grown out of metal manufacturing. For example, English Electric (aircraft and locomotive); Pressed Steel which made the Prestcold fridge (car production); and Tube Investments (bicycles). Others grew from electrical engineering, and by the 1970s and 1980s a handful of very large, multi-divisional companies, from both backgrounds, were competing in the market place, struggling to survive against an upsurge of highly successful foreign companies based in Germany, Holland, France, Spain, Italy and Japan.

Electrical engineering and appliance companies had, in fact, emerged in numerous countries following the pattern established by the USA. Although such giants as Sweden's Electrolux and Holland's Philips were established between the Wars, many others were products of the post Second World War years, particularly those in Japan such as Toshiba and Sanyo and Italy, such as Zanussi which have become international household names since the 1970s. These companies grew out of the belated industrial revolutions and consumer booms of their respective countries.

The Japanese consumer electrical goods revolution was for the most part, associated with audio equipment but, since the 1950s it has moved into the area of domestic appliances too. At first this was restricted to small appliances because the Japanese depended upon their export trade and therefore found it unprofitable to produce washing-machines and fridges. In addition, the traditional Japanese home was very small and there was a limit to the number of electric gadgets that could be crammed into it. Among the goods made for the domestic market — by such companies as Hitachi, Toshiba, National and Sanyo — were automatic rice-cookers and miniature ovens. The microwave oven was developed in Japan, becoming the first major manufacturer and exporter after 1960. One of the Japanese companies playing an international rôle in this area is Sharp. Characteristically, it had been a metal manufacturing company around the First World War and had moved into radios and televisions before making the final leap into domestic appliances.

Today, the domestic appliances industry is an important part of world trade, and companies from many countries produce goods across a wide spectrum of the market. While the Japanese manufacturers, for instance, concentrate on producing high technology goods at low prices, other — such as Sweden's Husqvarna and Germany's Gaggenau — manufacture products which are expensive but of an extremely high standard of workmanship and design. Because the world market for these goods is now so enormous, mass-production is the norm everywhere and the companies which have survived are, on the whole, large, and internationally orientated.

A British Electrolux delivery van from the 1930s.

4

NEW ENERGY, NEW MATERIALS

THROUGHOUT this century, the expanding market for household appliances has necessitated cheaper and cheaper mass-production and this in turn has stimulated research of a specifically technological nature. In fact, from the discovery of electricity to the development of the new, cheaper production techniques and materials for appliances, the industry has been highly dependent on technology.

Technology has provided an essential backcloth to the world of domestic appliances determining not only their presence, but also, their very nature and appearance. For instance, without the availability of steel and aluminium, many kitchen appliances would still be made out of heavy cast-iron. Similarly, the application of plastics technology to appliances after the Second World War brought a new, colourful look to many kitchen machines.

The roots of the technological background lie in the availability of alternative power sources to coal in the latter half of the 19th century — gas and electricity. Although the use of gas as a source of power for light-

A portrait of Thomas E. Edison.

ing goes back as far as the 17th century, it was not until the early decades of the 19th century that the industry supplied larger towns with lighting. While the first application of gas lighting was in factories and office and public buildings, electricity found its chief outlets in locomotives and lighthouses. It was some time, however, before gas and electricity became widely available to the domestic market.

The discovery of the incandescent lamp, in the 1870s by Swan in England and Edison in the USA, became more publicly significant with the improvements made to the dynamo.

This made it possible for electricity to become available for domestic lighting and by the 1870s it was a major competitor to gas. The Electric (Lighting) Act of 1882 consolidated its pre-eminence and the gas companies responded by promoting their power source as the best one for cooking. In 1882–3, the Crystal Palace

International Electric and Gas Exhibition strongly emphasized the advantages of gas in this area and companies hired out appliances to extend the use of their power source beyond that of lighting alone.

At the end of the 19th century, electricity was both expensive (in the 1880s bulbs cost 25 shillings and current was about one shilling per unit) and unevenly distributed. Its domestic use depended on the availability of current which, in turn, depended upon a number of private companies whose methods were largely incompatible. Inevitably, only the most populated urban areas had electricity supplied to them and it was not until the passing of the Electricity (Supply) Act of 1926 that efforts were made to standardize the British electricity supply system. By 1933 the original national grid was completed and 32 per cent of the country was connected into the electricity supply system as opposed to the 18 per cent which had benefitted from it in 1926. By 1938 this figure had reached 65 per cent and in 1945 it had risen to 86 per cent, demonstrating that it was not until after the Second World War that a large majority of the British population could use electrical appliances in its homes. Even then it should be stated, it was only irons and fires that accounted for most of the electricity used.

The situation in the USA was rather different. Although still essentially urban, a higher proportion of the population was connected to a central electricity supply by the 1920s. The manufacturers of appliances were more aggressive in marketing their products through advertising and other methods, such as employing Mrs Frederick to sing the praises of their products. A shiny, all-electric house, full of the latest gadgets, was shown at Chicago's Century of Progress Exhibition of 1933. This propaganda strategy had already been used in England at the Ideal Home exhibitions in the 1920s and the Empire Exhibition at Wembley in 1924, only at the latter show the building concerned was in 'Californian-Spanish' style. Also in Britain, a number of promotional projects were instigated in the late 1920s and 1930s, included an all-electric 'Tricity' restaurant in the Strand and the Carreras cigar factory in London. They served to spread the word concerning the advantages of using electricity as, indeed, did the work of the Electrical Association for Women formed in 1924.

Except in the areas of cooking and heating, electricity dominated gas, becoming the most widely used power source of all. The technological advances associated with both the invention of, and the widespread distribution of, electricity, had great bearing upon the evolution of domestic appliances. For instance, those appliances associated with heat – used for heating itself, cooking and ironing – were made possible through the discovery of the 'resistance' principle, whereby an electric current was sent through a substance which resisted it, so producing heat. In most cases this was created with the assistance of a long thin wire known as a heating element. The primary development, though, was the emergence of the electric motor in the late 1880s. The motor led to the development of a vast range of powered appliances, from fans to vacuum cleaners.

The electric motor was essentially a development of the work of Faraday in the 1830s. It used the same kind of equipment that Faraday had developed to provide power and light to convert electrical energy into mechanical energy. In 1864, the

N. Tesla's experimental induction motor, 1887/8.

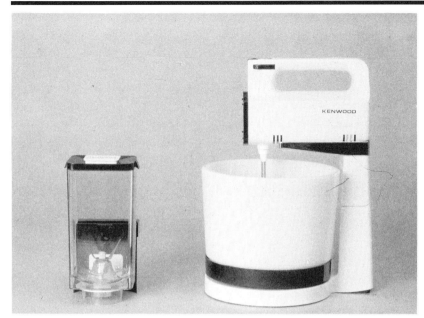

Kenwood's food-mixer designed by Kenneth Grange, 1965.

Dobbie Forbes and Company's cast-iron kitchen range, c.1895.

such as food mixers. An early solution in the USA to the problem of fitting motors to items of kitchen equipment was the 'kitchen power table'. This was equipped with a quarter horse-power electric motor on a lower shelf which was connected, by enclosed gears, to a power head positioned on the surface of the table. The appliances to be operated were connected to the pulley on the axle of the motor, or to one of the two horizontal shafts projecting from the power head. There was a slot in the top of the table through which various appliances could be securely fixed. Among the utensils to be powered in this way were, it was suggested, a bread mixer, a cake mixer, an egg-beater and a meat grinder. In addition, the power table could be used to operate a washing-machine, wringer, mangle, knife-sharpener, silver-polisher, coffee-grinder, food chopper, ice-cream freezer and vegetable slicer. By keeping the motor external to the objects it powered, it served a multitude of purposes.

Englishman James Clerk Maxwell extended Faraday's ideas in such a way that it led to the subsequent emergence of the electric motor. The first example was designed in the USA by N. Tesla and applied to a small table fan. The Westinghouse Company developed Tesla's work and, in the 1890s, made a number of significant advances on the basis of it.

The first motors were sold independently and could be attached to such products as sewing-machines and wringers to replace hand power. It was, however, to be some time before motors were small enough to become integrated into small items

With the imminent arrival of smaller motors, however, it soon became the norm for the motor to be integrated inside the casing. Such was the case, for instance, with food mixers, manufactured around 1920. The inclusion of the motor within the body dictated the dimensions of the product.

The early electric vacuum cleaners were little more than a sheet of metal curved to enclose a motor, fan and brush, with a stick and bag attached. The motors were housed in body-shells made of cast-iron, a material which substituted wood. The ability to make iron out of iron-ore went back centuries, but in the mid-19th century it became cheaper to manufacture than ever before and, as a result, more widely available. Its application in the construction of ships, bridges and the railways was enormous and in many ways Britain owes its rapid industrialization to iron.

In the early days, the product most dependent upon developments in cast-iron technology, was the kitchen range. In addition, all the implements that accompanied it – the grate, pokers, kettles and pans – were made of that same heavy, black material. The open range had originally replaced the open fire-place in the early 18th century and cast-iron panels were used from the 1750s onwards. Cast-iron baking ovens were soon fitted into the range and improvements continued to be made throughout the 19th century. After the 1850s, the range ceased to be an exclusively middle-class object and entered the working-class home as well.

The more efficient closed range, first advocated by Count Rumford, emerged for the first time in the early years of the 19th century and it was in the USA that range production reached its greatest heights. Giedion explained that, "*No country produced cast-iron stoves and ranges in such profuse variety as America.*"

Slowly, between 1850 and 1880, the gas range began to take over from earlier wood– or coal-fired models. Their forms relied heavily upon those of their predecessors but gradually the free-standing gas cooker, with its flat top and oven underneath, complete with curved, cast-iron legs (inspired by the cabriole legs of furniture) took over from the range. The dominance of cast-iron in the last century, made housework highly labour-intensive as the range had to be black-leaded every day and cleaning cast-iron cooking utensils was extremely arduous.

The evolution of the modern cooker is an illuminating case-study where the introduction of new materials – among them sheet steel and both porcelain and vitreous enamel — into domestic appliances is concerned. The increased availability (due to the discovery of the Bessemer process) of steel from the 1850s radically modified the appearance and production methods of all the large, cabinet-based domestic appliances. Once again, the USA led the way in these developments.

The USA's steel production expanded in the years between 1860 and 1930, helped by the war between the North and the South necessitating the production of guns, shells and railroads. Steel, together with iron, was a key to America's prosperity at that time and the vital ingredient in the production of its arms, sewing machines, bicycles and automobiles. The Ford Motor Company, for example, needed such large supplies of steel that it manufactured its own and, as Susan Strasser has pointed out, "*Easier to transport, rolled steel conformed better to turn of the century systems of centralized production and national distribution making for a general switchover from locally produced cast-iron to centrally produced rolled steel in American industry.*" By 1914 the American companies were responsible for a third of the world's total supply of steel. By 1940, however, most of the world's iron and steel was produced in Germany and, as a result, that country achieved pre-eminence in the production of metal consumer goods in the years following the Second World War.

The process of working steel that developed within the automobile industry was highly influential in domestic appliance manufacture. By bending and pressing steel, curved forms were created, replacing the heavier cast-iron antecedents and before long, cookers, refrigerators and washing-machines emerged, cased in all-steel body shells. At first, these were still attached to wooden, iron and steel frames, but gradually steel-pressing reached such levels of sophistication that no inner frames were needed.

Norman Bel Geddes' design for a gas stove for the Standard Gas Equipment Company of 1931 illustrates the way in which early advances in this production technology were achieved. Bel Geddes himself explained that, "*When I undertook to design stoves, only one or two manufacturers had discarded the custom of constructing the gas range, in part, of cast-iron*". His solution sought to eliminate areas which attracted grease and dust and to provide a flush ivory-white enamel surface, with just a touch of chromium plating on the handles. His stove had no legs and all its components were standardized for ease of mass-manufacture. He eliminated cast-iron once and for all and used sheet-metal construction throughout. The greatest problem that beset him was how to prevent damage to the steel sheets during transit and he solved this by attaching them to an independent steel frame which would take the shock if the stove was dropped. His other innovation was to replace bolt technology by welding, thereby eliminating much handwork as well.

Bel Geddes' preoccupations with the need for hygiene, rational production and easy transportation resulted in a design which provided a model for any others to follow. His solution combined ideas derived from, on the one hand, automobile manufacture, and, on the other, skyscraper construction and it succeeded in furthering the progress of the modern domestic appliance.

By the later 1940s, the vogue for streamlining dominated the world of domestic appliances but this too was the result, not simply of changing public taste but of technological innovation. As Harold Van Doren explained in an article in *Design* magazine in 1949, "*The truth is that much so-called streamlining is imposed on the designer by the necessity of obtaining low-cost through high-speed production.*" He used the example of the refrigerator to explain the main changes in manufacturing technology that had taken place during the course of the century. Until the 1930s, the ice-box, whether made of wood or sheet-steel wrapped around from front to back, predominated. This was changed by the work of the industrial designer Lurelle Guild, appointed consultant designer to the Norge company in 1929. His cabinet was produced in sections which were fastened on to a skeletonized frame, like Bel Geddes' stove. In terms of metal fabrication,

however, bending at sharp angles was replaced by making panels on large steel stamping presses – making mass-production highly economical. This method eliminated sharp corners and encouraged the popularity of streamlining.

A subsequent innovation, initiated by Westinghouse, was the use of what was called a 'tangent bender', which meant making refrigerator shells out of one long piece of 'U'-shaped steel and no inner frame was required. The radii of the curves on the corners could not be less than three and a half inches, thereby sustaining the vogue for streamlined forms. In 1940, the Philco Company went into refrigerator production using this process and applied the same principle to the manufacture of small appliances, among them toasters and vacuum cleaners. Whereas, for instance, pre-war American toasters had been made in three parts, a number of post-war models simply consisted of an inverted steel dome mounted on to a plastic base. As Van Doren explained, in this instance, "*Contours have to be soft in order to draw with the minimum number of operations and to permit of easy buffing and polishing, for these are among the most highly paid skills in the American manufacturing plant.*"

While iron and steel were the most important 19th and 20th century metals to influence the appearance of modern domestic appliances, a number of others, in particular stainless steel and aluminium, also played a vital rôle in their evolution. The twin advantages of aluminium were its lightness and non-corrosive properties and it became a popular material for a number of non-powered kitchen appliances, in particular saucepans and kettle. Following the War it was used for interior components of such appliances as washing-machines and vacuum cleaners, substituting heavier metals and thereby 'lightening the load of the housewife'. Often it was used as a motor housing and in a few isolated instances, as, for example, in Christian Barman's portable electric fire for HMV in the 1930s, its radiation properties were exploited.

Stainless steel had the same 'resistance to corrosion' advantages as aluminium. It remained, however, a more expensive material and was only used where there was no alternative or where its aesthetic properties enhanced the value of the object. A number of electric kettles and coffee-percolators exploited the luxury finish of stainless steel following the Second World War, replacing chromium-plated steel.

After metals, the material which has had the greatest impact upon the design of modern domestic appliances is the one described by the generic term 'plastics'. Just as iron replaced wood as a basic material, plastics have provided a cheaper substitute for metal on a number of occasions. Like steel stamping, plastics manufacture necessitates expensive tooling so it was not until the inter-War period in the USA that appliance manufaturers had enough capital behind them to be able to take advantage of the full potential of this new material.

The earliest plastics, among them celluloid and casein, had emerged in the USA, Britain and Germany in the second half of the 19th century but it was not until the discovery of a substance named 'bakelite' by Leo Baekeland in 1909, that appliances found a material that they could exploit. It was first used, in the late 1920s, by the American General Electric Company for the control knob on an iron but it was quickly picked up as the most suitable material to replace wood on the handles of kettles and irons. The earliest products to have body-shells made entirely of bakelite were telephones, radio receivers and cameras but the concept of the plastic body-shell for domestic appliances was still some way off as metal was the only material to provide all the necessary practical requirements. In his book *Plastics and Industrial Design* of 1945, John Gloag illustrated a number of plastic products – from jewellery to cutlery to furniture – but no appliances were included. Gradually, however, metal and plastic began to be used in combination with each other more widely. Toasters were given phenolic bases and more and more appliance details — oven door handles, the superstructures of irons, etc. — began to appear in plastic. While, for instance, Christian Barman's streamlined iron for HMV from the mid-1930s was made of glazed fire-clay and metal, the company's post-war model, the 4a controlled heat iron, had a black plastic handle and thermostat control knob mounted on to its enamelled steel base. A little later again, the 'IC10' model had a 'cool grip' black plastic handle mounted on to a plastic base which in turn fitted on to the smaller steel base. Where larger appliances, such as refrigerators, were concerned plastic was restricted to such details as the ice-freezing cups. In their book of 1941, entitled *Plastics*, V. E. Yarsley and E. G. Couzens predicted, however, that:

"Every small article with which man provides himself for his necessity or luxury, and which does not require metal for a cutting edge, or for electrical purposes or for heat resistance, will be plastic, and where metal parts are needed, as in a razor, safety or electric, or a sewing machine or an electrical device, all the casings and

solid structures which carry the metal parts will be moulded plastic."

Within a quarter of a decade, their prediction was to become a reality, and a wide range of electrical domestic appliances – from food mixers to vacuum cleaners to electric shavers — emerged with plastic body-shells.

One particular electrical appliance which took some time evolving an 'all-plastic' version, though, was the electric kettle. When polypropylene was launched in the 1950s, some thoughts were directed at applying it to the kettle but it was not until considerably later, the late 1970s in fact, that the plastic kettle finally emerged. The application of this new material led, in turn, to a radical new concept – the jug kettle – a tall, thin object which had more in common with the traditional coffee-pot than with the squat kettle. In 1976–7, Hoover moulded a one-piece handle and lid in plastic and attached it to the metal base of its electric kettle. Like the iron 30 years earlier, ease of manufacture resulted from this particular combination. An all-plastic kettle provided a special problem for the plastics, however, because while the thermosetting plastic, phenolic, had long been used for moulding housings and handles for electrical appliances, it could not withstand the constant transition

from cold to boiling demanded by a kettle. The final choice, by Hoover, of a plastic material called Kemata acetal copolymeer proved highly successful and was used by Russell Hobbs for its 'Futura' kettle – the first all-plastic model on the market. The numerous jug kettles which emerged after 1979 – the year in which Redring produced its revolutionary new shape – also depended upon the same material.

Plastics clearly represent the future for the manufacturers of domestic appliances and a number of Japanese companies have already produced all-plastic washing-machines and spin-dryers. More plastic details have also, in recent years, appeared on major appliances such as ovens. The justifications are, as with the use of new metal technology, decreased costs and ease of manufacture. With the new micro-processing

Jug kettle manufactured by Redring Electric Limited, 1979.

Russell Hobb's automatic kettle, 1956.

techniques used in manufacturing, the flexibility of using plastics increases. Materials and production technology move, inevitably, hand in hand.

The rôle of technological innovation in the evolution of domestic appliances has functioned, therefore, on a number of levels. Without researches into the application of new power sources, materials and production techniques, the story of modern appliances would have been very different. The final application of technological research concerns innovations in the mechanisms of the products themselves. Too numerous to list here, these include such inventions as the application of the thermostat to cookers and irons and the automatic functioning of toasters, kettles and washing-machines. Each modern appliance is, in fact, the result of a set of technological breakthroughs. Some of the more important ones will be discussed in more detail in the second half of this book in relation to individual appliances.

HMV's controlled heat iron designed by Christian Barman, 1936.

5

THE APPLIANCE DESIGNERS

"Modern design entered the American home not through the front door but by way of the kitchen, bathroom and garage."

W.D. Teague

THE question of the design of domestic appliances is the central theme of this book and the one into which all the other themes ultimately converge. Like their rapport with technology, the relationship between design and appliances functions on a number of different levels. Its most obvious relationship focuses on the concept of 'added value' — on the idea that the only way of expanding the market for appliances and of making people replace their old models with new ones is to make the products look desirable. Design is a commercial necessity. It is also the means through which object symbolism is injected

into appliances and by which they communicate meanings to their consumers and users. Design is, therefore, fundamental to the significance of domestic appliances in everyday life in this century.

The changing appearance of domestic appliances in the years since 1900 has largely been in the hands of a single individual — the industrial designer. He is the catalyst bringing together and interpreting, for popular consumption, the social, economic and technological constraints which have dictated many features of the modern appliance and he has translated them into forms with which we have become familiar. The designer's task — whether he be an anonymous in-house employee, a member of the engineering team with special responsibility for appearance, or a consultant brought in for a particular project — is a creative one. As a

free-lance, however, he didn't come into prominence until the late 1920s.

The look of the very early electrical appliances was determined first and foremost by utilitarian constraints, primary among them the level of available production technology. Early appliance manufacture, undertaken by the small, light engineering firms which moved into this area, was largely a one-off, or at most a batch production, process. The products were visibly the results of the crude metal bending process that created them. Hoover's first suction cleaner, as we have seen, was essentially a simple product of this kind. Interestingly, though, Hoover was also aware of the selling power of 'art' and added some Art Nouveau-inspired graphic motifs to the surface of the metal body-shell in the manner of the early sewing-machines decorated to conform to what the manu-

facturers called 'women's tastes'. Other early products, such as the 'American Beauty' iron, made no such concessions to the tastes of the market. Many early products – among them irons and coffee-percolators – simply borrowed their forms from their non-powered antecedents, often substituting more expensive materials and adding electric sockets. Decoration was used wherever appropriate, stamped, for example, into the metal sides of toasters and adorning early electric fires. Products which were intended for the living-room were much more likely to be adorned in this manner than kitchen-bound artefacts.

Generally speaking, though, in the years between 1900 and 1930 little effort was made to create a new image for electrical appliances. They sold instead on the basis of their resemblance to traditional objects — hence the coffee urn looking like a silver samovar. There was no need as yet to persuade consumers of their desirability because their very novelty provided a perfectly adequate selling point. Cookers were, for the most part, hybrid objects with metal cabriole legs supporting crude metal boxes; early refrigerators looked like wooden

'Model O' vacuum cleaner manufactured by Hoover Limited in 1908.

cabinets; electric frying pans like saucepans; and kettles like tea-pots. Objects such as toasters and vacuum cleaners, which had no real precedents to fall back on, had to create their own look as best they could. A strong sense of visual eclecticism characterized this early period of appliance design, or 'proto-design', and there were no signs of the emergence of a special, modern aesthetic for these technological products.

The domestic appliance industry first brought in an outside consultant in Germany to advise on the visual nature of its products. It was not simply to apply superficial decoration to his appliances, however, that Paul Jordan of AEG invited the designer Peter Behrens to work for him. He wanted Behrens to provide an entire visual character for the company, from its graphics, to its products, to the factories themselves. Behrens had had experience in the areas of graphic design, the decorative arts, interior design and architecture before working for AEG and the task of applying his skills to electrical appliances presented him with a new challenge. In many ways the results he achieved in his arc lamps, fans and kettles, were less than radical since he did little more, in the case of the arc lamps, than modify their basic proportions and he simply turned tea-pots into mass-produced electric kettles by

adding a socket. His designs provided, nonetheless, an early example of an artistically-trained designer putting his mind to the problem of the appearance of the products of the electrical appliance industry.

What Behrens did for AEG's products was to show that they were best left undecorated and that as simple products of a new industry, they should incorporate an aesthetic all of their own. The seeds of a commitment to a 'machine aesthetic' not fully penetrating the world of appliances until considerably later were already evident in Behren's work. This sense of commitment was best expressed by Paul Jordan when he said, "*don't think that even an engineer, when he buys a motor, takes it to bits to scrutinize it. Even he as a specialist buys from the external appearance. A motor ought to look like a birthday present.*"

This idea was best put into practice in the USA between 1930 and 1960 when selling products became so crucial to the industry. At this time the consultant industrial designer emerged to take control over the world of consumer goods, particularly those emanating from the 'new industries' as they were most in need of his attention. The reasons were complex but mainly derived from the special economic situation in the USA in the late 1920s. The consumer boom of the earlier decade came to an abrupt end in 1929 due to the Crash and as a result, the large manufacturing companies, which had expanded rapidly since the close of the First World War, suddenly felt their markets contract dramatically. The crisis experienced by Henry Ford in 1927, forced him to go back on his commitment to a single, standardized unchanging product. Instead, he adopted General Motors' principle of the 'annual model change'. This quickly spread to other industries — particularly those involved with domestic appliance manufacture, many of which had links with the automotive industry. By the late 1920s, 'visual expendability' in products had become commonplace. All the designers brought in to cope with this *volte-face* had backgrounds in two-dimensional design and advertising and easily conformed to the needs of the manufacturer who, in desperation, called upon their specialist skills.

The first large-scale company to confront the question of design in its products was General Electric. Like General Motors, GE was a decentralized company and its home

Peter Behrens' kettles for AEG: 1909 and 1910.

appliances section was entirely self-contained. In the 1920s, GE turned its attention from the technical to the aesthetic content of its goods. In 1912 the company had employed a J. W. Gosling who was responsible for the appearance of its goods. However, such design didn't really become a specialized activity in the company until 1928, the year in which Ray Patten was brought in as a designer for GE's home appliances. A special unit was created and, gradually, design work became increasingly specialized, resulting in design groups assigned to specific product lines. The design teams were responsible for the entire design process from the creation of clay or plastic models through to full-scale prototypes and production models.

The product which aroused the greatest attention in the USA, *vis-à-vis* its visual appearance in the inter-war years, was the refrigerator, which became the second most important status symbol after the automobile. The General Electric Company played an important rôle in the design changes in that product during those years. For instance, the 'Monitor Top' fridge, first produced in 1926, was an example of a product

'Model 150' vacuum cleaner manufactured by Hoover Limited and designed by Henry Dreyfuss, 1936.

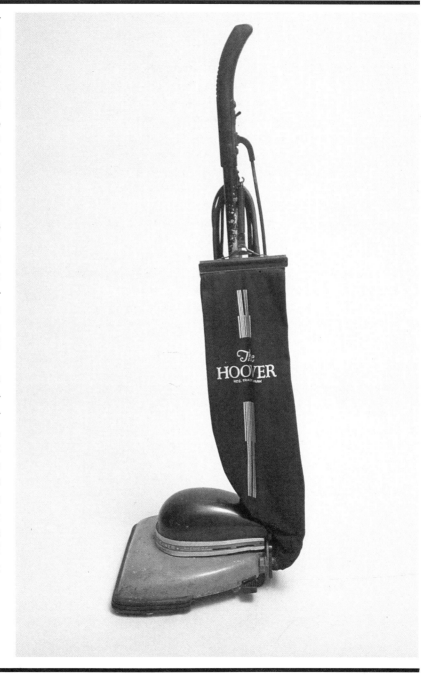

49

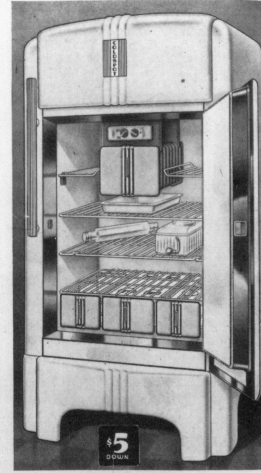

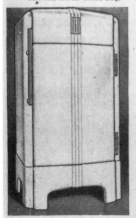

COLDSPOT

REG. U.S. PAT. OFF.

"Super Six"

**Lovely Modern Design
Super-powered "Package Unit"
Full 6-cubic foot size
About half usual price**

A NEW COLDSPOT for 1935 and a NEW Standard of Value in electric Refrigerators. By Value we don't mean just a lower price. You will never appreciate the Value offered in this COLDSPOT merely by looking at its price. Here is all we ask: Forget the price for the moment and consider this COLDSPOT purely in terms of Quality. Study its Beauty. Check its features. Analyze it strictly in terms of what it offers you. Then compare it with any other refrigerator of similar size, selling in the $250 to $350 class. We say that you will find the COLDSPOT actually a *Better* refrigerator, *In spite of the Fact That It Costs Only About Half as Much.*

USE YOUR CREDIT. You don't have to pay cash. See Easy Payments Prices and Terms on page at right.

All Prices for Mail Orders Only.

VEGETABLE FRESHENER
Large, covered, porcelain enamel vegetable freshener for keeping lettuce, celery, tomatoes, etc. in a fresh, crisp condition. Easy to keep clean and sanitary. Slides in and out exactly like a drawer.

STORAGE BASKET
Large wire basket, containing two over-size covered glass dishes to keep butter, salads or left-overs from absorbing the taste of other foods in the box. These dishes can be removed for kitchen use if desired.

STORAGE BASKET
An open wire basket for holding coarse vegetables, fruits, etc. to eliminate breakage. (This container and the 2 shown at left suspend from lower shelf like drawers.)

WATER COOLER
Covered glass water cooler with down faucet. Holds about a quart of liquid. Can be used for iced tea, lemonade or other beverages. Especially desirable during the hot months.

576   prices in this **Sears** catalog are for mail orders only

---

Raymond Loewy's 'Coldspot' refrigerator designed for Sears Roebuck in 1935.

whose appearance was determined more by utility and technological necessity than by aesthetic idealism. As Don Wallance explained in his book *Shaping America's Products*, "*the cooling unit of this model was frankly exposed where it could operate at maximum efficiency...Even the apron at the bottom of the cabinet, originally intended to conceal a pan for collecting drippings from the ice, was retained in this model.*" Although yearly modifications were made to the design, its appearance remained fairly static. By the mid-1930s, it had become clear that when compared with other models available on the market — notably those manufactured by Frigidaire and Sears Roebuck — its appearance had become very old-fashioned.

At this time General Electric, and indeed the other fridge manufacturers as well, went to the well-known, highly successful, consultant industrial designers of the day to seek advice. GE first consulted Norman Bel Geddes but this collaboration

Right: W.D. Teague's streamlined electric oven for Floyd Wells Company, mid-1930s.

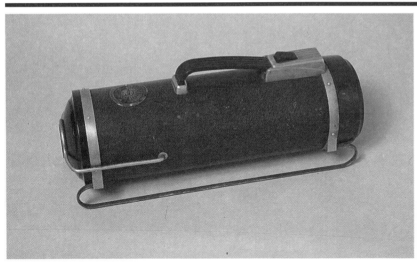

Left: Electrolux Limited. Cylinder vacuum cleaner, 1930s.

Below: Electrolux Limited. Cylinder vacuum cleaner, 1950s, designed by Sixten Sason.

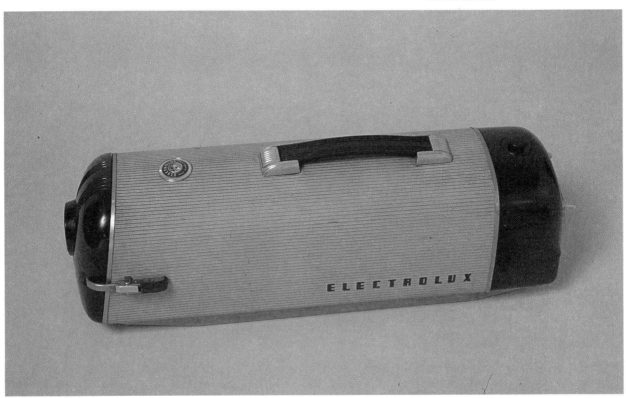

reaped no results so they went to Henry Dreyfuss who in 1933, was retained on this particular project. He worked closely with the GE staff — both designers and engineers — and his new model, launched in 1935, eliminated the cabriole legs of the earlier fridge. He also developed a completely different range of components integrating the sides with the top, brought the cooling unit inside the body and made such details as the hinges and handles much less cumbersome. The following year, this elegant new refrigerator was replaced by a new model which had a more bulbous, streamlined front in line with the current fashion for that particular style. In 1939 it was modified yet again with the introduction of ridges on the front, enhancing its association with the 'Streamlined Moderne' style which depended upon ideas about aerodynamics being translated into a new aesthetic, primarily for objects of transport. As Ray Patten explained, "*the radii of cabinet corners and edges became larger, front doors swelled outwards and the tops became more dome-like in an effort to make the refrigerator appear larger and more imposing in visual competition with other makers on the dealer's floor.*"

The same tendency could be observed in automative design in these years and in the appearance of other appliances. The most notable

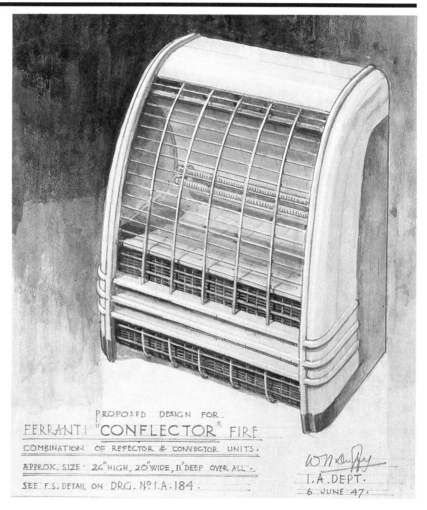

Electric bar fire designed for Ferranti Limited by W.N. Duffy, 1947.

examples included Raymond Loewy's 'Coldspot' fridge for Sears Roebuck which he first remodelled in 1935 and provided annual model changes in 1936, 1937 and 1938. From 1939 onwards, Loewy was retained by Frigidaire to work on a range of kitchen appliances including a fridge, range, automatic washer, dryer and iron. In 1930, Westinghouse, too, employed an 'in-house' designer – Donald Dohner – and in the same year the company launched its first refrigerator.

While the refrigerator companies were very quick to realize the benefits of making their products desirable, competitive objects, they were not alone in the appliance field. Dreyfuss, for instance, also designed Sears Roebuck's 'Toperator' washing-machine in 1933, and in 1936 he worked on what was perhaps his best known project, the Hoover 'model 150' vacuum cleaner. Hoover was beginning to feel pressure from competitors in the mid-1930s. Electrolux had made a significant impact on the American market with its cylinder machines and Bel Geddes' 1934 model, and Montgomery Ward and Singer were also making vacuum cleaners. However, the visually powerful model that Dreyfuss designed for Hoover quickly brought the company back as a brand leader. It remained a familiar model until well after the Second World War. Walter Dorwin Teague, often referred to as the 'Dean of industrial design', also worked on a number of appliances in the 1930s, including Crosley's 'Shelvador' fridge of 1935 which was the first of its kind to have shelves in the door.

Where small appliances were concerned, examples such as Lurelle Guild's 1934 food mixer for GE and Egmont Arens' 'Kitchenmaid' mixer for the Hobart Manufacturing Company, produced in 1949, showed that streamlining was as important a style in minor as in major appliances. Countless irons, toasters and coffee-percolators were designed in the same idiom in these years. In his book *Industrial Design: A Practical Guide*, Van Doren explained how the industrial designer went about designing a toaster in the 1930s. He points out that it was not simply a question of improving the appearance by eliminating clumsiness, but also of looking very carefully at competing products.

The term 'streamlining' was replaced by one critic with the alternative 'cleanlining' and new products were frequently sold on the basis that there were no longer any dirty, dust-collecting areas. The inter-war years' near obsession with hygiene meant that appliances were nearly all white or just off-white. Whatever its significance for the housewife, the main function of the streamlined look for the manufacturer, was 'added value' at point of sale. As the 1930s progressed, the housings became increasingly swollen and chrome details more expressive as a means of turning products into their own best advertisements. Donald Dohner explained in 1936 that,

"a few years ago several leading electric refrigerators were designed not from the standpoint of appeal or convenience, but from that of tradition, ease of manufacture and possible freedom from field troubles. The simple fact that an electric refrigerator is an improvement over the old ice-box should have suggested a slight break with tradition, leading to finer appearance."

After the Second World War, the debt that household appliances owed to automative design in the USA became even more explicit. As Adrian Forty has explained, *"Given that cars were the first machines with which people became familiar and that they were the first mechanical goods to receive close attention by stylists, it was not surprising that domestic appliance design tended to imitate the appearance of cars."* Also important was the 'status symbol' argument which encouraged people to think of appliances as they did about their cars. As the kitchen became part of the social area in the house and the housewife turned into a hostess it became increasingly important for kitchen appliances to become objects of 'conspicuous consumption'.

The most obvious debts to car design were the use of automobile-style door-handles on fridges and the addition of heraldic-type emblems on the fronts of appliances and the resemblance of control panels on ovens to those in cars. By the 1950s, colour suddenly entered the area of appliances, blending with the new live-in kitchen which had first appeared before the War and by the middle of the decade consumers could match their appliances with the colour schemes of their interior decor. This helped turn the kitchen from a laboratory into a social centre. Frigidaire was the first company to launch its refrigerator in a range of colours. By the end of the decade, however, the all-white appliance had regained its pre-eminence and sharp corners had replaced the curved ones.

The USA led the way in appliance design right through from the late 1920s to the late 1950s. What little design went on in other countries — such as the work of Sixten Sason in Sweden and Christian Barman and Douglas Scott in Great Britain — tended to reflect US innovations.

In Germany, however, an alternative aesthetic emerged in the decade after the War. It ousted the USA quite quickly as a world style-leader. German models were more scientific, rational and restrained, respecting the laws of simple geometry and classical proportions. The style evolved into a very precise aesthetic by the early 1960s relying upon the minimum of visual means.

This 'machine aesthetic', as it came to be called, was much more in line with the tenets of 'good design' being formulated by many countries after the Second World War. It was epitomized in the products which emerged from the Braun company, but quickly came to characterize the goods of numerous other German appliance manufacturers as well. In Britain, the work of the consultant designer, Kenneth Grange, for Kenwood, Morphy Richards and a number of other companies, owed much to the German machine style and introduced a new purism into British product design.

A square look dominated appliance design through the 1960s and 1970s, contrasting violently with the expressiveness of American streamlining of the previous decade. It was an aesthetic which had grown, not out of the commercial world of appliance manufacture and sales *per se*, but from the more high-minded design ideals formulated by the British Arts and Crafts movement and transmitted via the Bauhaus and Modern architecture into the area of domestic appliances. By the early 1970s, however, the concept of the 'black' or 'white box' had become ubiquitous as a synonym for modern domestic appliances and it looked as if the ultimate product forms had been achieved. It conformed also to the idea, expressed by Henry Ford much earlier in the century, that the best products were standardized ones

Braun space heater H1, 1959.

manufactured in restricted colours. Standardized mass-production and centralization of industry marked the 1960s and 1970s.

The result was that appliances became bland, neutral accompani-ments to living rather than the exciting status symbols they had been in the USA in the 1950s.

With the oil crisis and the demise of the unfaltering belief in the power of high technology in the second half of the 1970s, it became increasingly apparent that the 'square' look was symbolically inadequate. There were isolated attempts at evolving a Post-modern look for appliances but few went into production due to the industry's need for capital investment in tooling and its reluctance, as a result, to experiment with a new style. In Italy, in 1980, for instance, the young Post-modern designer, Michele de Lucchi proposed a set of small appliances' which took their aesthetic from children's toys. His maquettes appeared in a range of pastel colours and unusual shapes. In England, a young designer called James Dyson proposed a design for a Post-modern vacuum cleaner employing similar colours and looking utterly unlike the minimal machines it would compete with on the market.

The most exciting venture into new forms for domestic appliances came, however, in the early 1980s from Japan. With the advantage of a young, affluent, fashion-conscious home market to cater for, the

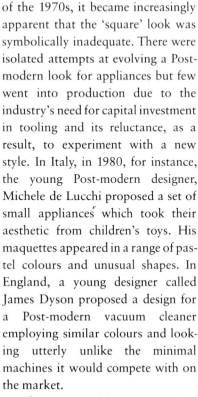

Left: GEC coffee percolator designed by Douglas Scott for Raymond Loewy Associates, late 1930s.

Right: Maquette for vacuum cleaner designed by Michele de Lucchi for Girmi in 1979.

Japanese appliance manufacturers were in a much stronger position to experiment with new shapes and colours and quite quickly Sharp, Toshiba, Hitachi and others had launched ranges of Post-modern appliances which totally rejected the minimal, cool look of the previous decade. Pinks, turquoises and other vibrant colours were employed and the objects made frequent visual references to the American appliances of the 1950s from which they had taken their stylistic inspiration. In 1986, National Panasonic launched a range of small appliances, even more dramatic in their use of new colours and forms. The flexibility of Japanese production methods made it possible for them to experiment in this way.

The future of appliance design is less than clear, but it seems that the concept of 'life-style' is a more important determinant now than technological expediency. The recession of the 1980s has, in some ways, paralleled the 1930s and it seems, therefore, that appliance designers will become increasingly conscious of the rôle their products will play in providing a symbolic backcloth for the lives of their consumers.

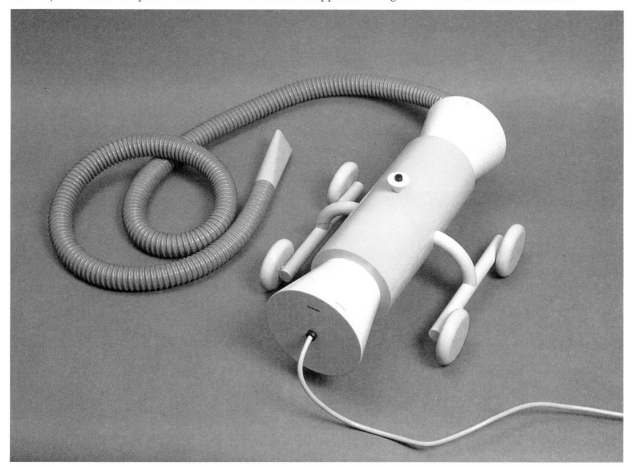

PART TWO

6

PREPARING
AND PRESERVING FOOD

COOKING is amongst the oldest of all the everyday domestic activities. Ever since the discovery of fire, people have transformed food from a raw into a cooked commodity and its preparation has become an important household task. Cooking over an open fire, whether on a spit or in a pot suspended from a hook, remained the main means of preparing food for domestic meals until the 18th and 19th centuries when first the open and then the closed range made it possible to put utensils on an iron surface covering the heat source. Enclosed ovens, particularly for baking bread, accompanied the arrival of the range making the spit and hanging cooking-pot redundant.

In the second half of the 19th century, the substitution of coal for gas as a source of heat provided the first major transformation in domestic cooking. It took about 30 years, how-ever — from 1850 to 1880 — for the transformation to catch on, moving gradually from hotels into the home. By the end of the century, wealthier homes often had gas ovens in their kitchens in addition to coal-fired ranges in the fire-places, retained as a means of heating water for the household. The gas range was used primarily in the summer when the solid range became too hot.

The 19th-century gas cooker was liberated from the fire-place becoming a free-standing object. It was this that provided the visual model for the modern electric cooker. The competition between gas and electric cooking has remained a constant through this century, especially in Britain where the price of gas made it highly competitive. It was not until after the Second World War that cooking by electricity – the price, relative to gas, had decreased substantially – became a significant activity in Britain. In the USA, however, gas was less strong a competitor in the early years and electric cooking became the norm there at a much earlier date. Commercial electric cookers were fairly widely available in the USA by 1910. After the Second World War, however, a shortage developed due to cessation of production during the War and gas again became more popular for a short period of time.

The first British electric ovens, made by Crompton & Company, were demonstrated at the electrical exhibition at the Crystal Palace in 1891. Numerous other manufacturers quickly emulated their example, often converting cast-iron gas stoves into electric versions. Some were made of sheet metal but they still had a heavy black surface with large obtrusive door handles, hinges and oven controls — the very first ones, in fact, resembled safes. Electric cookers were often smaller than gas ovens and were raised up on little cabriole legs. In larger households, though, bigger, double-doored versions were frequently found. While the main electric switches remained on the walls, the control switches – usually fabricated from porcelain enamel – were mounted on to the front of the cookers. These early examples were considered by many to be, and often were, quite dangerous and difficult to use as oven temperatures were impossible to control. One of the key factors behind the retention of the gas cooker was the servants' fear of using these 'new-fangled' electric objects.

Standardization in cooker size was totally absent in these years and as a result varied enormously. In the USA, the split level cooker, with a hob on one side and a raised oven on the other, became popular but few kitchens in Britain could accommodate its bulk. Where colour was concerned the all-black cooker was gradually replaced by versions which had mottled, vitreous enamelled panels as well as some all-white enamelled panels.

By the 1930s, however, more white components gradually appeared on electric cookers and,

by the middle of that decade, the 'all-white table-top' cooker had appeared in the USA.

A number of technological improvements were made to the electric cooker through the middle years of this century. While some of the more extreme experiments, such as linking the cooker with the refrigerator proved unsuccessful, others, such as the introduction of interchangeable boiling plates, removable interiors, glass doors,

radiant boiling rings, and, eventually, thermostats, were all essential discoveries in the evolution of the modern electric cooker. Early experiments in heat regulation had been undertaken by Crompton & Company and GEC around the turn of the century, and early ovens had thermometers on the front to show the heat obtained. It was not until 1931, though, and several years after its introduction into the gas cooker, that a thermostat for the electric cooker

Split level 'Carnifix' oven and hob by AEG, 1930.

Kitchen with split level oven and hob, AEG, 1986.

was finally perfected. In Britain, Creda was the first manufacturer to introduce one into its 'Credastat' model.

The electric cooker undoubtedly reduced the amount of labour involved in cooking domestic meals. A daily task for the servant who used a solid fuel range had been both lighting the fire and black-leading the stove. The use of the new power source and the introduction of

Advertisement for Hotpoint table-top range, 1955.

enamelling eliminated both these tasks and made cooking cleaner and easier. Before the invention of the thermostat, however, cooking had to be constantly supervised.

In spite of the obvious 'cleanliness' advantages of electric cooking, only six per cent of British homes had an electric cooker in 1935. This figure rose to 22 per cent in 1955 and 48 per cent in 1985. The reasons for this were essentially twofold: expense and mistrust. From the early days, electric cookers had been more expensive than their gas equivalents

and more uncontrollable. There were numerous exhortations from women's associations to move over to the new clean power source but public reticence to do so was much in evidence. One highly popular exception to this rule in Britain was Belling's little 'Modernette' cooker which appeared in 1919 as the solution to cooking in the small house or flat. This small, all-steel cooker was the first of a number of highly successful electric cookers manufactured by Belling. Others included a 100 per cent vitreous enamelled version in

1931 and the 'Vee' cooker, designed in 1945 for the post-War prefabricated temporary housing supplied by the Government.

Reservations about electric cooking were not so apparent in the USA. Electric ranges – as they were called there – were available from the turn of the century. General Electric's first model appeared, for instance, in 1905. Early American cookers resembled heavy solid-fuel ranges but they gradually became lighter as the years went on. Around 1927 choices of colour became available but until the mid-1930s, electric ranges had little legs and no storage space. The greatest American innovation around this time was General Electric's invention of the metal covered 'Calrod' surface unit replacing the use of bare wires, or coils of wires, embedded in porcelain brick.

The increasing popularity of electric cooking in the USA was a result, as elsewhere, of the electricity supply industry realizing its importance as a large consumer of current. After 1930 gas cookers had to improve their appearance in an attempt to be as attractive as their electric competitors. As Giedion has written: "*Now it* [the electric cooker] *has a gleaming white enamel casing and the oven cannot be distinguished from the utensil drawers. It has merged with the kitchen.*"

This new look for the cooker was

Sanyo microwave oven, 1980s.

largely the work of the American industrial designers who set about making the electric cooker part of the 'modern' domestic environment. American electric ranges remained considerably larger than their European counterparts and it was in the second half of the 1930s that the double, all-white, porcelain-enamelled, streamlined 'table-top' electric range, complete with thermostat and black bakelite handles first made its impact on the American market. In 1938 the Sears Roebuck catalogue illustrated its 'Prosperity' model. In 1952, according to a McCalls' magazine report on home appliances, white was still the preferred colour for electric ranges.

The 1950s was the decade in which the new emphasis upon the consumer led American manufacturers to add countless details to their appliances in an attempt to woo new customers. A Crosley electric range

for instance, illustrated in *Ladies Home Journal* in 1950, was described as combining 'beauty and brains' and came complete with an automatic oven timer, a 'beautiful chromium hooded lamp which lights the inside of your pans, a self-sealing oven door and seven heat speeds.' By the middle of the decade glamour had entered the American kitchen, brought in by the refrigerator but rapidly adopted by ranges as well. Frigidaire's models, for instance, came in a range of colours, including pale yellow and the 'glamorous new charcoal grey' and the control panels at the back came to look more and more like those in a contemporary automobile from Detroit.

The main challenge to the giant-scale, double-fronted American kitchen electric range came at the end of the decade with the decision to split the hob from the oven, thereby enabling baking to become as much an 'eye-level activity as frying and grilling. The 'split-level' concept quickly penetrated international cooking habits and became a popular alternative to the all-in-one cooker. Often a gas hob was combined with a raised electric oven.

Another innovation in domestic cooking which had penetrated a significant number of households by the 1980s, was the appearance of the miniature microwave oven. It was developed in the USA in the 1950s and exploited commercially through the 1960s and 1970s by a number of Japanese manufacturers – among them Sharp, Hitachi, and Sanyo. This radically new form of cooking became increasingly popular as a result of accelerated life-styles all over the western world and the accompanying increased need for fast cooking. It facilitated the rapid re-heating of pre-cooked food and generally made the cooking process a much faster and easier one.

By the 1980s the cooker market was dominated by the European giants – Bosch, Neff, Zanussi and, for the more style-conscious, affluent consumer, Gaggenau which took the 'minimal' cooker to its logic extreme. Split level models with such innovatory features as ceramic and halogen-heated ceramic hobs, and electronic controls and one-piece cookers were available either in brown, to match the fashion for 'rustic' kitchens — part of the general tendency for nostalgic styling in the '80s — or white, fitting in with the other popular style of the day epitomized by the 'high-tech' look. Kitchen accessories ranged from 'peasant'-inspired brown earthenware to stainless steel items suiting the more futuristic version of the domestic environment. The general mood, in whatever style was selected, was one of total integration and the cooker became simply one of the numerous pieces of

hardware making up the efficient fitted kitchen. While the popularity of fast food and microwave cooking in many ways decreased the need for the full-size cooker, it nonetheless continued to be an important feature in the modern kitchen.

IN the early days of electric cooking, the stove was only one of a number of other, more specialized, electrically powered cooking utensils, many of which were small and decorative enough to be brought into the dining-room. The 'all-electric kitchen' at the Chicago World's Fair of 1893 contained, for instance, an electric kettle – the first of its kind – and some electric hot-plates, or 'table stoves' as they were called then. In the first two decades of the 20th century, the presence of such items as toasters, coffee-percolating pots and urns, chafing dishes, small grills, hot plates and waffle irons was commonplace in affluent American homes. Don Fregnant claims, in fact, that "*the small kitchen appliance field was America's most active growth industry during 1890–1930.*"

Because of their frequent appearance on the dining-table, these small appliances had strong visual affinities with traditional serving items. Coffee pots, for instance, were little more than electrified samovars and the hot plates resembled 19th

century silver table services. Many of the objects – chafing dishes and coffee-percolators among them – had little legs, vestiges of the oil-fired hotplates on which they had previously sat. Such items remained very popular in the 1920s and 1930s and they were quickly joined by a number of new appliances, among them corn poppers, waterless cookers, sandwich grills, pancake griddles, doughnut bakers, hot dog cookers and egg boilers. Some were highly versatile combining such functions as boiling, steaming, frying and toasting in a single object. They all eliminated the necessity for one member of the family, usually the housewife, to stand over a hot kitchen range, while the rest sat in the dining-room. Other reasons for their popularity included their suitability for small kitchens and ability to cook small quantities of food as families became smaller. Gradually, however, with the return of the housewife to the kitchen after the Second World War, many versatile 'dining-room' appliances vanished, with the exception of only a few.

The story of the electric kettle is atypical of that of most electrical appliances because it was not dominated by American examples. Early cast-iron kettles boiled water by being suspended on a hook over an open fire but they were really only pots. The kettle with a spout, however, only dates back to when tea and coffee drinking was introduced into the western world in the 18th century. As a result of the growing fashion among a wide spectrum of society for tea-drinking, the kettle gradually appeared in more attractive metals, particularly copper and brass. The advent of electricity a century later made it possible to bring the kettle into the dining-room, first on little hot-plates and later with an electric heating element attached to the bottom of the kettle. As with most electrical appliances there was much public resistance to this new object, particularly as water and electricity were involved in such close proximity to each other.

The major technological kettle advances were made in Europe. It was in Germany, for instance, that the first kettle with an immersed element appeared. Peter Behrens' decorative little kettle was produced by AEG in 1909 and seems to be the first of its kind, in spite of the fact that recognition for this particular innovation is usually given to the English company, Bulpitt & Sons with their 'Swan' kettle of 1922. Before the Second World War electric kettles were often chromed or nickel-plated although the non-powered, aluminium version remained a favourite with many householders for a long time. In the USA, where tea-drinking was less popular than in Great Britain, electric kettles were relatively rare sights.

The greatest attention, where technological advances were concerned, was given to safety, focusing on the perennial problem of the kettle boiling dry. A number of different fail-safe mechanisms were introduced, most using a bi-metallic strip, but it wasn't until the mid-1950s that the Russell Hobbs Company perfected the automatic kettle which successfully switched off when it had boiled. This coincided with the predominance of stainless steel in kettles

A 1921 copper electric kettle with element in separate compartment.

and the influence of more stream-lined forms, making a final break with the traditional brass and copper kettles which had not been radically modified for over 100 years. Although they were independent of gas and electric ranges, kettles retained their flat bottoms and rounded forms for some time.

By the mid-1970s experiments were taking place with plastics, combined at first with metal but eventually replacing that material altogether. The most significant innovation of the 1980s was the appearance of the all-plastic jug kettle, finally breaking the link between the kettle and the habit of tea-making by making it closer in shape to the conventional coffee-pot. The reasons for

Right: Seymour-Powell's design for a cordless kettle for Tefal, 1986.

Below: Russell Hobb's 'Futura' kettle of 1976.

this change were essentially twofold: the jug shape was easier to work in plastic and it was part of an attempt, on the part of British kettle manufacturers, to expand their markets, moving into countries which did not have such a long-established tradition of tea-drinking as Great Britain. The experiment proved successful and the sales of jug kettles expanded quickly through the 1980s.

IF electric kettles developed in Great Britain predominantly as part and parcel of the social habit of tea-drinking, coffee-makers were largely an American invention and preoccupation. From the early days of the availability of mass-produced home appliances, coffee-making was high on the list of priorities. Like so many other electrified small appliances, the electrification of coffee-making meant that the householder was in control of the process and didn't have to wait for the servant to provide what was needed. At first this meant using a small hot plate but quite quickly, as with the kettle, a heating element was added to the base of the coffee-pot itself. In his article entitled 'Electricity in the Household' of

Right: Vacuum coffee machine manufactured by Cona Limited in the 1970s.

Left: Electric coffee machines manufactured by AEG between 1926 and 1961.

1890, A. E. Kennelly listed electric coffee-making alongside such other early electrified objects as the light, telephone, bell, door-opener, fan and phonograph explaining that:

"The advantages to a man whose duties call him out during the night, of being able, from his bedroom, to set an electric coffee-heater at work in his dining-room, so that by the time he is ready to leave the house he finds hot coffee awaiting him, and all without arousing any person in the house, far outweighs the three or four cents for electrical power that the beverage has probably cost him."

He went on to describe the coffee-maker in question as either an ornamental stove enclosing a coffee-pot, or a kettle in an asbestos lining with coils of wire around it. By 1914, electrified percolator urns and, a little later, percolator pots were available – the former with a central tap and two handles and the latter with a single spout and handle. In form, they were little more than traditional objects with an added base, while those with their own built-in elements retained elegant legs. Westinghouse described its model of these years as an 'electric percolator, colonial style'. It was decorative, nickel plated and intended to blend with the furnishings in the dining-room. The two alternative machines – the urn and the pot – remained in general use on American breakfast tables throughout the 1920s and 1930s. However, the percolator jug generally became more popular and the urn eventually disappeared from prominence.

In his book *The Electric Home* of 1934, the American writer E. S. Lincoln explained that by that time there were two main kinds of coffee-making machines available – the percolator and the metal or glass machine using the drip method. While an aluminium percolator was adequate for the kitchen, the silver-finished, glass or porcelain model was more appropriate in the dining room. Few, as yet, were automatic, so the use of safety fuses was advised

The models employing the drip method were either made of metal or consisted of two glass bowls fitting one on top of the other. The 'Silex' model was the most popular one in the USA in the 1930s: it sat on a small heating unit and could be converted into a tea-pot with the addition of a small chromium finished cap.

The electric coffee-percolator was introduced into Britain in the inter-war years but as coffee-drinking wasn't widely popular in that country, at least until the introduction of the instant variety in the 1960s, it never became the ubiquitous object that it was in the USA. Coffee had long been a drink for ordinary people in America. In her book *Never Done* Susan Strasser describes for example, how 19th century American farming families ate their own produce exclusively and spent what little extra they had on coffee. In Britain tea-drinking had, since the late 19th century, performed this same social rôle and coffee-drinking was reserved for more affluent families. Occasionally, non-powered glass coffee-makers, made by the Cona Coffee Machine Company, appeared in idealized images of the British kitchen after the Second World War but, in reality, it was an exception rather than the rule to own such a device.

The style of American coffee-percolators changed, after the Second World War, and sleeker, more streamlined forms replaced the 'reproduction' look of earlier models. By the 1950s automatic machines also became increasingly popular and models such as Universal's 'Coffee-Matic' and Sunbeam's 'Coffee-Master' graced countless American breakfast tables. Chrome-finished, and with black bakelite handles, these machines had only a simple engraved motif on their sides to show that they were as at home in the dining-room as in the kitchen. The 'Coffee-Matic' publicity boasted that

Left: Russell Hobbs' automatic coffee percolator. Model CP3, 1970s.

it brewed automatically, signalled when the coffee was ready and kept it at the perfect serving temperature. By 1960 an even simpler, Scandinavian-inspired elegance had entered American coffee-machine design: Westinghouse's spoutless coffee-maker was a prime example of this new, sculptured style.

In Europe the simple filter method of coffee-making had long been favoured and in Italy, in particular, the 'expresso' method – used at first exclusively in cafés and for mass-

Filter coffee machines by Braun Limited, 1970s.

catering — had become highly popular by the 1950s. By the 1960s and 1970s these European methods had become increasingly sophisticated and began to penetrate the domestic market as well.

A significant change in the evolution in home coffee-making after the Second World War was its return to the kitchen. With the final demise of servants and the growing involvement of the kitchen with general home life, the distinction between 'dining-room' and 'kitchen' appliances was eroded. The main effect of this change was upon the appearance of the products which became decreasingly decorative and increasingly utilitarian. By the 1970s, however, attempts to turn the kitchen into a living — as well as a working — space brought with it a neo-decorative approach to small appliances and such nostalgic motifs

GEC electric toaster of 1946.

as sprigs of corn and flower images appeared on coffee-percolators, kettles and toasters in an attempt to bring a rural atmosphere into the high-tech kitchen.

The complexity of the drip method and the danger of 'overcooking' associated with percolating coffee gave way, in the 1970s and 1980s, to the growing popularity of the European methods of making coffee. As their machines became cheaper, too, more people turned towards them. The affects of increased foreign travel were another important factor in this change. The most popular appliance to fulfil this need was the automatic filter coffee-machine, developed in Germany by such manufacturers as Rowenta, Siemens and Braun. In Britain, sales of filter-coffee machines had risen to 1,035 million by 1983, although instant coffee still accounted for about 90 per cent of all coffee drunk in that country. The predicted follow-up to the filter machine is the domestic expresso coffee-maker and many European manufacturers – among them Cimbali in Italy, Rowenta in Germany, Salton and Russell Hobbs in Great Britain and Philips in Holland are presently testing the market for that particular appliance. Both filter and expresso machines have opted for the neo-functional 'plastic box' appearance and are most definitely kitchen bound appliances. A few exceptions, however, such as Salton's aluminium expresso machine, have more in common with the sculptural Italian machines of the 1950s which brought the craze for coffee-drinking into Britain in the post-war years.

Another kitchen appliance which, like the kettle and the coffee-machine, goes from strength to strength is the electric toaster. Initially a British invention of the early 1890s this appliance was among the first domestic utensils to make use of the new power source, providing an alternative to the traditional toasting fork or iron toasters held in the open fire. Like the kettle, the electric toaster depended upon the principle of 'resistance wire' heating and it very quickly became a familiar item on the breakfast tables of the more affluent households in the first decades of this century. Also like the kettle and the coffee-maker too it provided a means for the householder being able to make his own breakfast without having to rely on servants to prepare it. As Mrs Peel explained in 1917, "*You do not need to ring for more toast but make it yourself and eat it while it is crisp and hot.*"

Because the toaster was intended to grace the dining-room table it was quick to adopt a decorative appearance and to reject its early manifestation as a rather dangerous looking collection of metal wires. In 1914 an American writer explained how the early models worked: "*The heating element of the toaster is a series of bare coils of thin wire. These coils are placed in a horizontal position behind a wire screen. The bread to be toasted is placed on this screen. The*

current heats the wire coils and the heat passes up to the bread by convection and radiation."

Later models had nickel- , and a little later, chromium-plated steel sides with patterns cut into them, a means both of avoiding over-heating and of providing decoration. A number of modifications were made enabling toast to be easily turned and thereby toasted on both sides and racks were put on the top to keep the toast hot until needed. Although the search for the automatic 'pop-up' toaster was intensified in the years following the First World War, the non-automatic version still remained very popular, particularly in Britain where it was a familiar object in many homes in the 1950s.

The invention of a toaster which automatically shut off when the toast was ready was American in origin. The Sunbeam 'Toastmaster' of 1926 was the first example and the advertisement which heralded it boasted "*Pop! up comes the toast automatically when it's done and the current is automatically turned off. The toast is made in a jiffy because both sides are toasted at the same time. There is no guesswork.*"

Even in the USA the automatic toaster only gradually replaced the hand-operated model and in the mid-'30s there was still a range of options available. Many resembled examples from the early century with ornately pierced sides, while others conformed to the call for more modern streamlined forms influencing so many of that decade's appliances. Sears offered a variety of styles from their 'De Luxe Automatic' to their 'Oven-type' and 'Moderne' models, the last one having an Art Deco pattern cut into its side panels. The most advanced American model from the 1930s was, undoubtedly, the 'Toastmaster' which, with its boxed form, resembled models to come. The convex panelling on its side served both to strengthen its steel sides and to create an attractive sculptural appearance.

The 'open element' type (rarely automatic) and the 'oven' type (frequently automatic, helped by a thermostat controlling a clock mechanism) vied with each other for supremacy until the Second World War after which, in the USA, the latter leapt ahead in the popularity stakes. As a book about appliances of 1934 explained: *"with automatic types the possibility of burning the toast is eliminated and there is the added convenience of not having to give the toaster your constant attention."*

After the War, the sides of American toasters became increasingly bulbous and streamlined. The 1950s were dominated by such forms, the toaster body now pressed out of a single sheet of steel and covered with shiny chromium and boasting controls which allowed for light or dark toast according to preference. In a copy of *Ladies Home Journal* in 1950, the actress Shelley Winters sang the praises of the Knapp Monarch model and in 1955, Universal launched its 'Toastamatic' which controlled the colour of the toast. New construction techniques in the later 1950s and 1960s brought with them, however, the return of the sharp cornered 'box', which has remained the norm until today.

The first British toasters to emulate American examples was produced by Morphy Richards in 1956. It came with a yellow enamelled body and quickly became a very popular model — in 1957 100 per cent of British consumers who chose an automatic toaster bought the Morphy Richards' product. In 1961, the company replaced its first streamlined toaster with a new model which was both squarer and cheaper. Malcolm J. Brookes explained in *Design* magazine that *"Whereas the*

Morphy Richards' electric toaster with heat adjuster, c.1965.

Russell Hobbs' sideways toaster, 1970s.

casing of the old model has the typically large radius corners and rounded forms of pressed steel, the new model has the look of a precision instrument with crisp, clean forms, black Bakelite ends and a polished chromium cover."

Through the 1970s and 1980s toasters retained this look, essentially remaining metal boxes encasing the mechanism within. Gradually, following German examples, the boxes have become slimmer and more elongated in an attempt to reduce bulk and, as lip-service to the move away from the ideal of the kitchen as a laboratory to the 'live-in' kitchen, patterns and motifs have appeared on the sides of many models. Countless European manufacturers, from Philips to Siemens, to Braun to Moulinex, have moved into toaster production, producing a wide range of models which compete with each other in the market place.

The toaster has become a ubiquitous appendage of modern living and, like the kettle and coffee-maker which it so often accompanies, it is now a kitchen bound object for the

HMV 'Horizontal' toaster, 1946.

most part, vying for a position on the increasingly crowded work surfaces.

Siemens. 'Universal Toaster TT 5237', 1980s.

Although not involved with cooking itself, a very important kitchen appliance where the preparation of food for cooking is concerned is the food-mixer or processor. Traditionally, the work involved with food preparation involved innumerable small tasks including cutting vegetables, beating eggs and batters, kneading dough, grinding coffee, whisking cream and chopping meat. This in turn necessitated a vast number of food preparation utensils – spatulas, knives, egg whisks, chopping tools, coffee-grinders, apple-parers, etc. – which proliferated in the second half of the 19th century, especially in the USA, causing kitchen cupboards and containers to become increasingly crowded. Even in the 'organized' kitchens of the early advocates of sci-

entific management in the home, these tasks were undertaken with the help of a wide range of implements. Of course, they were grouped efficiently in the area where the task was performed. The concept of defining each task served, in fact to highlight the need for such single-purpose tools and encouraged the continued presence of traditional tools for the job of food preparation.

With the advent of the electric motor, however, and its application to the home, it became possible to provide another means of driving these tools besides mere manpower. The idea of the 'kitchen power table' has already been discussed. It proved an invaluable asset to many kitchens, providing a means of driving not only the bread mixer, cake mixer, egg beater and meat grinder but also the washing machine and the mangle. It constituted in fact, an interim stage in the evolution of the electrically-driven food mixer and processor which performed a number of food preparation tasks.

By the 1920s, a number of fairly primitive food mixers had appeared on the market, both in Europe and the USA, which performed a number of specific tasks and had electric motors integrated into their bodies. The first American model, produced by Landers, Frary and Clark in 1918 consisted of two beaters with a motor above them which, in turn, was fixed

to a stand. This arrangement remained essentially unchanged for several decades. Gradually other tasks, such as mincing and shredding, also came to be performed by mixers through the addition of the necessary attachments. Later, liquidisers, can-openers and juice-extractors were combined into this multi-purpose utensil and the highly complex food processor that emerged after the Second World War had become a reality.

As early as 1922, in *Cheating the Junk-Pile*, Ethel R. Peyser illustrated a complex-looking food preparer, produced by the Troy Metal Products Company. It plugged into the light socket and performed a number of tasks including coffee-grinding, meat-slicing, mayonnaise– cake– meringue– and batter-mixing and vegetable– and fruit-cutting and preparing. In addition it had a soup strainer and colander and an ice-cream freezer. Visually the machine consisted of a large motor attached to the top of a hood holding the various attachments. The motor was also fixed to a stand holding a mixing bowl.

It wasn't until the 1930s that electric motors became small and reliable enough to make the food preparer a more popular object. It also permitted the designer to integrate it into the main body of the product, inside a metal casing. By the middle of the

Portable electric food mixer, c.1920.

decade, the 'Kitchenaid' machine had become a much more unified object, its motor sheathed in a body-shell attached to a stand. Where production was concerned, however, the product was still a complex object, fabricated as it was from numerous components. Although now rationalized in appearance, it still had to acquire the aggressive image that was to follow.

In the 1930s many large food machines came with their own containing cabinets so that all the accessories could be stored efficiently. These 'complete food preparing outfits', as they were called, were sold alongside smaller, more specialized, electric mixers, beaters and juice extractors, drink-mixers and ice-cream freezers. Post-Second World War American food-mixers and preparers were among the most stylized of contemporary domestic electrical appliances. Back in 1935 Sears' 'Power-Master' had already indicated this tendency and by the 1950s, so had a wide range of manufacturers – including Sunbeam (whose 'Mixmaster' machine constituted 34.9 per cent of all sales in 1959), Dormeyer, Hamilton Beach, Westinghouse, Kitchenaid, Montgomery Ward and Universal. In 1950 the Hamilton Beach model came with see-through pyrex bowls and the 'Mixmaster', with its curved handle arched across the back of its body and streamlined motor casing, showed a huge debt to contemporary automobile styling. This was reinforced by the description of its 'Hi-speed mixers'. The 'Kitchenmaid' model of the same year offered 'exclusive planetary mixing action' to ensure uniform mixing.

By the end of the 1950s, as with so many other products, the emphasis had moved away from expressive shapes and towards minimal crisp forms. The main reason for this change derived from the contribution of German appliance-styling in these years and food mixers and processors were no exception to this rule.

The machine that introduced an entirely new way of thinking, where food preparation was concerned, was Braun's 'Kitchen Machine', produced in the late 1950s. Food preparing machines had long been inspired by industrial machinery and many food mixers had closely resembled such objects as routers and drills. Braun took this idea even further, introducing the industrial metaphor of 'rationalism' into the appearance of its food-mixer. The 'Kitchen Machine' conformed to Braun's general house style, one which Richard Moss has described as being characterized:

'Mixmaster' food mixer, manufactured by Sunbeam in the 1950s.

"by the use of white, the weighty masses, the purity of lines, the company signature and so on. There are other resemblances, more subtle and more powerful, which derive from the fact that all Braun products are designed in the same spirit... three general rules seem to govern every Braun design – a rule of order, a rule of harmony, and a rule of economy."

He went on to note the use of parallels in the horizontal surfaces, edges and joints of the all-purpose 'Kitchen Machine'. It was also among the first food processors to be fabricated predominantly in plastic.

The Braun 'Kitchen Machine' inspired a number of other European food mixer designs, notable among them Kenneth Grange's 'Chef' manufactured by Kenwood. Formed in 1947 the Kenwood company's first 'Chef' food-mixer of the mid-1950s had resembled American models such as the 'Kitchenmaid' machine. It was produced alongside a large, commercial model intended for use in cafés, hotels and hospitals and a smaller, portable food mixer, which had much in common with Sunbeam's model.

One of the ploys Kenwood utilized to market its first Chef model was to supply a nose-cap, cover-cap and control switches in a range of five 'gay' colours intended to 'match the new look in your kitchen'. The influence of the American 'glamour' kitchen of the 1950s was clearly being felt on English soil and Kenwood was quick to exploit all the marketing implications of this fad.

The move from coloured plastic accessories to the strong visual formalism of Grange's early 1960s redesign of the Kenwood 'Chef' marked a transition from the USA to Germany as the country which had the greatest influence on the design of British domestic electrical appliances. Kenwood brought Grange in as a consultant designer to the company and succeeded in producing a food-processor which was to remain in production for many years with only minor modifications. The original design came with a metal body covered in white stove enamel with plastic relief in blue. It was smaller and lighter than its predecessor and both mincing and liquidising accessories could be attached to it. The new 'Chef' proved to be an enormous success and remains one of Britain's most noteworthy designs in this field.

In the 1970s and 1980s, the European market for food mixers and preparers became increasingly dominated by the French and German food processors which provided a multitude of functions. With fewer people making bread and cakes at home the emphasis increasingly

Range of Kenwood 'Chef' food mixers – 1950s, 1964 and 1980s models.

Braun's 'kitchen-machine', 1959.

moved from mixing as the main task, to the preparation of vegetables, and a number of products emerged, the best known manufactured by Moulinex, Rowenta and Braun, which were particularly adept at chopping and shredding vegetables. With this change came a new image

Moulinex food processor, 1980s.

for the food processor. The dominance of the bowl and whisks was replaced by a drum into which the vegetables were introduced.

A tendency of recent years has been for a number of companies to move away, in a pendulum swing, from the idea of the multi-purpose machine, back to simpler, single purpose tools. The reason for this is twofold; one is because the complex machines are often too difficult to assemble and clean for simple tasks and the other is the fear that once a major machine has been bought, the consumer will not be prepared to invest in anything else. Thus a number of small powered whisks, mashers, tin openers and other such items have emerged recently in the hopes that the graph of consumption will not begin to fall – a stimulus which has inspired much innovation in this area through this century.

Numerous other small kitchen appliances — electric frying pans, sandwich-makers and yoghurt-makers among them – have always existed alongside the more prominent ones. Some fall into the area of 'gadgets' while others have reduced labour considerably. It is an area which will undoubtedly continue to expand, not least because it is easy to export small items and their relatively low price makes them highly 'consumable'. While some, such as the 'chafing dish' have faded from

Above: Leonard, Moreton and Company's electric frying-pan, 1980s.

Right: Toast-maker by Tefal, model 39198, 1980s.

Far right: Philips' ice-cream maker, 1980s.

view, others, such as the domestic expresso coffee machine are increasing in popularity. Inevitably they follow fashions in eating habits and life-styles and the more lasting

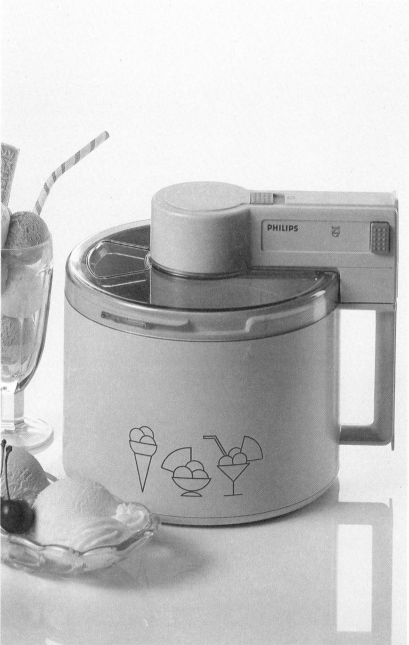

among them – in particular the kettle and the toaster – provide familiar items in nearly every modern household.

"THE *chic porcelain-lined refrigerator of today is the corner-stone of the halls of domesticity.*"

Ethel R Peyser 1922

Unlike the kettle, but like the coffee-maker, electric refrigeration was an American invention and development, and all the major advances in the nature and appearance of the electric refrigerator have taken place in the USA in this century. While one explanation for this is the hot, humid summers experienced by so much of the USA, the other is linked to the inaccessibility of shops for so many rural Americans. This meant that both the home produced and purchased foodstuffs had to be preserved as carefully as possible and refrigeration was therefore a basic necessity for many American households.

The origins of the modern 'fridge' can be found in the ice-box, an ubiquitous artefact in 19th-century American homes used to keep food cool. Its efficient functioning depended upon a regular delivery of fresh ice and, for this reason, the ice-box stood, for many years, on the back porch where it was easily accessible to the ice-man.

The need for mechanical refrigeration increased in the second half of the 19th century. As Ruth Cowan Schwartz explained: "*between 1830 and 1880 dozens of mechanised refrigerating machines were patented – machines that would make ice as well as machines that would cool large compartments without making ice.*" The two principles – compression and absorption – which were to be perfected later as the twin bases of modern refrigeration, were, in fact, discovered in this period but it was to be some time before they were made widely available. Inevitably, like so many other innovations of this

'Arctic' ice-safe, 1920s.

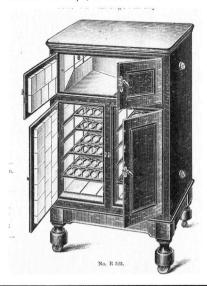

No. R 323.

period, the mechanical production of ice was used commercially in beer-making and meat storage, before it entered the domestic arena. Before the advent of the electric motor, steam engines were used to provide the necessary mechanical power.

The search for a domestic mechanical refrigerator was a relatively lengthy one as it involved a number of complex technological developments which took some time to perfect. All mechanical refrigeration utilizes the vaporization and condensation of a 'refrigerant' as a means of creating low temperatures. When this process is powered by electricity it involves a component called a 'compressor'; by gas, however, it operates on the principle of 'absorption'. The story of the development of the modern refrigerator in the years between 1900 and 1930 in the USA focused on finding a suitable refrigerant – in 1931, Frigidaire discovered 'Freon' which replaced sulphur dioxide. It was also about asserting the pre-eminence of electricity over gas – compression over absorption. As Ruth Cowan Schwartz has pointed out, however, the dominance of the former over the latter was less one of natural superiority than of the large manufacturers opting for electricity and therefore forcing it on the market.

There is continued debate about the identity of the first domestic

highly significant 'consumer durable', playing an important role in the American economy of those years. It also became a focus for many of the design ideas in the air at that time, receiving the attention of many of America's leading industrial designers.

The first major innovation was Frigidaire's decision, in 1926, to produce a steel refrigerator. However, its product still looked much like the wooden cabinet it replaced. It wasn't until the emergence of General Electric's 'Monitor Top' model of the following year, which put the compressor on top in order to reduce bulk, that the compact refrigerator, with integrated machinery and box, was seen as a real possibility.

The collaboration of the large manufacturing companies – Frigidaire, General Electric, Kelvinator, Westinghouse and Sears Roebuck – with the designers through the 1930s led to the appearance of the streamlined refrigerator which, like the automobile, became a major American status symbol in that decade. In addition Loewy's 'Coldspot' introduced the idea of stylistic obsolescence, borrowed directly from automobile manufacture and which served the same purpose in both products — to encourage and

refrigerator – contenders include the 'Domelre', first manufactured in Chicago in 1913; the 'Kelvinator' manufactured by Goss and Copeland in 1918; and the 'Guardian' model, a wooden electric refrigerator first produced in 1915. The company which produced the 'Guardian' worked on a small-scale basis and quickly ran into financial trouble. It was bought, as a result, by General Motors in 1918 when searching for a new post-war product. The first 'Frigidaire', still a wooden model, was produced in 1918.

These early refrigerating cabinets were far from perfected, technologically. The key developments took place between 1923 and 1933, coinciding with the increased wealth of the American market and the mass-production potential of the large manufacturers. By the 1930s the electric refrigerator had become a

increase consumption. From its late start in the domestic context, the electric refrigerator became a necessity in the majority of American homes, and by the 1920s the idea had emerged that every household was a potential customer for this new product. This was a result, largely, of huge capital investment in mass-production technology, huge advertising campaigns and public relations exercises.

A number of design innovations also took place in the 1930s which moved into the post-War years. Among them were the Crosley Corporation's decision to use the inside of the door for extra shelving, the introduction of stainless steel for

Frigidaire refrigerator of 1955.

shelves and the foot opener for users whose hands were too full. In addition, the use of automotive styling and chrome trim turned the refrigerator into a highly desirable appliance. It was, however, after the Second World War that the American refrigerator blossomed fully into the 'dream machine' so envied by other countries all over the world, inspiring style exercises in many of them.

General Motors' Frigidaire division dominated all American developments in refrigerator design in the 1950s influencing, in turn, the appearance of other kitchen machines. While there had been a brief period, at the end of the 1920s, when refrigerators were still expensive and were sometimes painted to suit a particular decor, through the 1930s, 1940s and early 1950s, the all-white refrigerator became a gleaming symbol of health and efficiency. In 1952 the refrigerator was voted the most popular appliance in the USA. With the new post-war interest in the kitchen-living space and the housewife acquiring, with the help of her 'labour-saving' goods, the image of a glamorous hostess rather than that of a servant substitute, refrigerators, pioneered the way for domestic machines to become as much decorative and symbolic items as they were functional. This was achieved exclusively through the use of colour and in 1955, Frigidaire offered a range in

International Harvester Company: Home freezer, 1952.

'Sherwood Green', 'Stratford Yellow' and 'Snowy White'. Back in 1952, a report in a magazine entitled *My Kitchen* had claimed that the most popular kitchens of the day had either light yellow or green walls. A Frigidaire advertisement in *Ladies Home Journal* of 1955 explained that: *"glamour comes into the kitchen with appliances in colors to match walls and cabinets."*

Another Frigidaire first of the mid-1950s, was a joint fridge-freezer. By 1957, Frigidaire had introduced what it called the 'sheer look' and the bulbous refrigerator became, as a result, a thing of the past. The company described its new model as 'straight, smooth, flat to fit' and for the first time the refrigerator became integrated into the fitted kitchen. This tendency was reinforced in 1960 with the 'sculptured sheer look' which now came in a wide range of 16 models and sizes, in white, 'Mayfair Pink, Sunny Yellow, Turquoise, or Aztec Copper', complete with an abstract motif printed on the door. The automobile imagery which had dominated American refrigerator design since the mid-1930s had finally given way to an image fitting in with the general decor of the new 'live-in' kitchen environment.

The other American post-war innovation in home food preservation was the domestic freezer which rapidly grew in popularity, particularly in rural areas. Dubbed the 'fair-haired boy among post-War appliances', the potential market for this new product paralleled that of the refrigerator in the 1920s. American manufacturers were quick to seize this opportunity and in the years after 1945 both upright and chest types became widely available. Frigidaire designed its upright model along the same lines as its refrigerator and offered its combined freezer-refrigerator as an alternative.

American refrigerators dominated the European market as well as their own up until the 1950s. For example, when owning a refrigerator became increasingly popular in Britain in the 1930s, General Electric's 'Monitor Top' was among the most popular models available. Another alternative, however, was the

Swedish company, Electrolux's gas powered 'absorption model' which remained a viable alternative in Britain until after the War. Electrolux opened its British factory in Britain in 1927 and began a programme of aggressive marketing. It took much longer, however, for the British population than it did their American equivalents, to be persuaded of the benefits of owning a refrigerator. While only 114,700 refrigerators were sold in Britain in 1948, by 1959 sales had increased to 750,000.

When the streamlined fridge gave way to the more geometric 'built-in' model in the late 1950s, the focus moved away from the USA and towards the domestic production of more rational 'unit' refrigerators. A notable British example was Prestcold's 'Packaway' fridge which received a Design Council award on the basis of its "*finely radiused corners, the precision of the assembly, the slight profile in the matt anodised aluminium cover strip, the well shaped handle and its constraining chromium finish, the neat lettering of the name-plate, the absence of self-conscious 'streamlining' and 'styling' coupled with almost classical proportions.*"

This sense of geometry remained a characteristic of the numerous refrigerators that were produced in Germany and Italy through the 1960s and 1970s. It was during these years

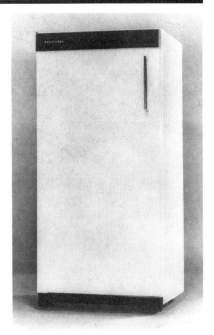

Prestcold's 'Packaway' refrigerator of 1960.

also that the fridge-freezer became a popular European machine. Where design was concerned, anonymity and self-effacement dominated the 'white boxes' that emerged from the European factories. The refrigerator ceased, as a result, to be a symbol of overt affluence and became instead, like the ice-box it had superceded, a functional machine with few claims to individualism.

7

LAUNDERING CLOTHES

"The mechanized washing machine is as typical and natural a product of America and full mechanization as the precision watch is of Switzerland and highly skilled handicraft."

S. Giedion.

WHEN the home was mechanized in the second half of the 19th century in Europe and the USA, laundering was an exception. While the activities of sewing, canning and preserving food, for example were commercialized and taken out of the home, washing clothes became – with the development of machinery to facilitate it – a chore increasingly associated with the domestic sphere.

By the mid-19th century it had become common practice in many European countries for washing to be done either in commercial laundries or public wash houses. Increasingly, from the 1860s middle-class, American households, too, adopted the habit of sending their washing to laundresses or commercial laundries. Catherine Beecher proposed this, in fact, as a much better way, claiming that washing clothes should be excluded from household work. Commercial washing and ironing also became a means of economic survival for many unmarried women, unable to find other forms of work.

In the USA, especially in the poorer households, much washing still went on at home remaining one of the most arduous of all household tasks. As a result, the need to find a means of 'saving labour' when washing began to preoccupy countless inventive American minds after 1860. Thousands of solutions to the problem were in fact put forward before one preferred model of mechanized clothes washing began to emerge. While the commercial laundries utilized steam as a means of forcing dirt out of clothing, the domestic machine was based upon the traditional 'dolly' principle and ways were sought of eliminating hand operations from the process.

Early machines were simply wooden tubs placed on a stand with a set of beaters or paddles which, driven by a flywheel, agitated the clothes. Sometimes a wringer was attached to the tub. Inevitably the earliest models often damaged clothing through rough action and they frequently leaked. Gradually, however, they were improved and by 1910 were almost as popular as sewing machines. The 'agitator principle' remained the norm for most domestic machines, even when electric motors were added in the first decade of this century. Among the first companies to launch an electric washing machine were the Hurley Company introducing the 'Thor' model in 1907; A. J. Fisher, who entered the market a year later, and the Maytag Company which launched its electric machine in 1911. In the years up until 1927 (the year in which Maytag manufactured its millionth washing-machine), the consumption graph for electric washing machines in the USA rose dramatically and it became clear that it was a product with a future. Like that of refrigerators, washing machine production benefitted from the consumer boom of the 1920s and the new manufacturing capacities of the larger companies.

Few radical advances had been made with the appearance of these machines, however. The exception was the substitution of wood by steel, either galvanized or porcelain enamelled. The washing-machine was still essentially a round tub or drum on legs with a motor visibly attached to it. Maytag was the first company to square-off the corners of the drum, aligning it more directly with the other major kitchen appliances, but most manufacturers retained the round form.

An important factor in the visual conservatism of the washing-machine was the ambiguity of the preferred location for it. While 19th-century washing had taken place in the scullery or, where such a room was available, in the laundry, with the smaller houses and apartments of the 1920s, it inevitably found its way into the kitchen. Therefore, unlike

the range, and, increasingly, the refrigerator, the washing-machine was an uncomfortable addition to the kitchen and it took some time to

adapt its form to its new location. In addition, most advances in washing-machine design and technology now occurred in the commercial sphere

Opening at London School of Electrical Domestic Science showing Principal demonstrating an electrical washing and mangling apparatus in the 1930s.

before they went on to influence domestic models. As a result, the latter adopted the overtly industrial appearance of the former, failing to acquire an identity of its own.

Several different methods of washing clothes by electrical means emerged in the inter-war years alongside that of the 'dolly' – among them the 'rotary' or 'cylinder' method, the 'oscillating' method, the 'vacuum' method or a combination of all three. Wringers were also electrically powered and worked in reverse action to ensure that maximum water was expelled from clothing. By the late 1930s, a few companies – among them Maytag and Sears — had succeeded in 'styling' their products in such a way that they looked less like old-fashioned washtubs with a motor attached and more like pieces of modern equipment. The copper drum was encased in a white, or cream, porcelain enamelled and squared-off steel cabinet, also encasing the motor. The wringer was also shrouded in a casing which helped turn the product into a 20th-century appliance. Soon the legs were to vanish and the washing-machine became a unified steel box, much more in sympathy with its neighbours, the range and refrigerator.

The most dramatic advance in the evolution of the modern washing-machine came in the USA at the end

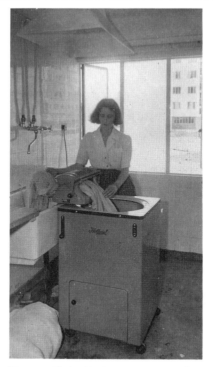

Housewife in 1948 in a communal wash-room in Walthamstow's 'vertical village' with Hotpoint washing machine.

of the 1930s, with the advent of the fully automatic machine. This was not to penetrate the market in any significant way, however, until the decade after the Second World War when it became one of the most desirable appliances available. Although it was an expensive item, more and more consumers aspired to owning one of the new Bendix, Maytag, Frigidaire or Westinghouse machines which filled, washed, spun and

drained clothing without any help from the housewife. The improvements on earlier and contemporary cheaper models were enormous. The problems of filling the machine, taking wet clothes out and transferring them to a spin drier or passing them through a wringer were now a thing of the past.

Back in the 1930s a British household manual had described the workings of the only type of electric washing machine available at that time: "*The clothes are put into the already prepared soapy water, the switch is turned on and at the end of 15 or 20 minutes the first batch of clothes is removed and another put into the boiler*", demonstrating just how time-consuming the early models had been. Now, with automation, the housewife could be doing something else while the washing was being done. The new American machines of the early 1950s were described in advertisements as having 'gyrafoam' and 'live water' action and they undoubtedly eliminated the concept of 'wash day' once and for all.

The impetus for change had again originated in the commercial sector. Bendix – originally an aeronautical company which had expanded during the Second World War and needing a new product to take it into the post-war period – came to prominence by supplying front loading machines for the new American self-service 'wash stations' before it entered the domestic market. It was Bendix which introduced the front-loader into the home, thereby providing a machine which could be easily integrated into a fitted kitchen, not needing to be moved to be used. The company also pioneered what was called the 'tumble action' method, replacing the long-standing pre-eminence of 'agitation' as the dominant washing-machine principle. Not all automatic machines were front-loaders, however. For instance, in the mid-1950s, Maytag perpetuated the principle of top-loading, producing, massive rectangular machines with 'Detroit'-inspired control panels at the rear. RCA's 'Whirlpool', washer-dryer combination machine of 1957 was similarly inspired. High prices prevented the entire population from possessing automatic machines, however, and the 'wringer-washer' remained popular in many homes in the USA through the 1950s, with the Maytag model dominating the market.

Alternative models to the fully automatic machine were particularly evident in Britain. The widespread use of powered washing-machines, as of so many other similar products, was slow in developing in comparison with the USA and with Germany.

The large, industrial American machine of the inter-war years found few outlets on British soil. Instead, hand washing methods and gas or electrically heated wash boilers were very popular. The latter were simply cylindrical machines heating the water in which clothes could be washed. Wringers were also used, often fixed over the sink. By the 1940s, boilers were available in a porcelain enamel finish. Because they were small, relatively cheap and heated water more efficiently than the early washing-machines – many of which had to be filled with hot suds – the boiler remained the dominant machine in British home laundering. After the War, the Hoover Company manufactured a very popular little washing-machine with a wringer attached, little more than a modernized, electric boiler. Rectang-

Zanussi automatic washing machine, 1980s.

Twin-tub washing machine by Hoover Limited, 1956.

ular in form, and with a fold-down position for the wringer, it was a neat little appliance which fitted easily into the post-war British kitchen. It was advertised as being 'easy to move and clean' and, it was claimed, it 'could be stored under the draining board'. One of the reasons why it was possible to introduce an effective washing-machine with a smaller drum was the introduction of powder detergents which created fewer suds than soap. After the Second World War, the problem of heating water to boiling point was also minimized by the widespread use of synthetic fabrics which could be washed in cooler temperatures.

A range of possible machine types were available in Great Britain in the 1950s, from the large American automatics – Bendix, Thor and Hotpoint predominantly – to the electric washers with manned wringers, the most popular of which were provided by Hoover and Goblin.

While washing-machines, particularly automatics, were the preserve of wealthy, middle-class consumers in Britain in the 1950s, the situation changed dramatically at the end of the decade. John Bloom began to import inexpensive models from Holland and since this was so successful he began manufacturing the same models in England. Craig Wood, the new managing director at Hotpoint also directed his attention to the cheaper end of the market and mass-produced a newly designed model retailing at an unprecedented low price. By the early 1960s, the washing-machine had become a ubiquitous object in British homes.

The most popular model was still not the fully automatic one however which was restricted, for the most part, to the newly-emerging chains of self-service 'launderettes'. It was the 'twin-tub' machine combining a washer and spin-drier in the same cabinet which was most in demand. This model had emerged in the USA in the 1930s and at that time simply consisted of two tubs joined together by a steel strap. Gradually the product was styled to become integrated into a single body shell. Inevitably, with the advent of the increasingly cheap fully automatic machine in the USA the 'twin-tub' gradually became redundant and the spin-drier ceased to be a separate unit. This could only be achieved once the 'agitator' principle had been superceded.

It wasn't until the 1970s that it was possible to talk about the widespread use of fully automatic machines in Britain. This was made possible by the rapid increase of European manufacturers in the area – from Zanussi in Italy, to Philips in Holland, to Bosch in Germany. The front-loading, white box dominated the picture in the 1970s and 1980s (although colours have recently been introduced). It continues to provide the most familiar image of the washing machine in the majority of British homes.

THE history of drying-machines inevitably parallels that of the washing-machine. In the USA in the 1920s, large drying cabinets, filled with hot air in which clothes were hung on racks, were the electrical equivalent of the traditional 'clothes horse'. Their bulk made them utterly impractical for use in the home, however, and the roller wringer whether manually operated or powered by electricity, remained the most popular form of eliminating water from wet clothing.

The spin drier was a German invention but it was picked up by the USA and sold as part of laundry equipment, either as a separate unit or attached to the wash-tub. Combination washer-driers were fairly common in the USA between the Wars but didn't penetrate the British market until considerably later. Inevitably, with advances in micro-technology, the tumble-drier has now been fully integrated into the automatic washing-machine and the presence of two large units in the kitchen or wash room will, undoubtedly, soon be a thing of the past.

THE iron is the only household appliance which takes its actual name from the material from which it was traditionally made. Ironing clothes with a flat piece of heated metal goes back for many centuries and the traditional 'flat' or 'sad' iron was widely used, often in pairs. While one was being heated on the fire or range, the other was being used to take creases out of newly washed clothing or household linen.

The electric iron is a product of the 1880s, in France first of all but quickly afterwards in the USA and Great Britain. Early models were cordless and had to be continually replaced on an electrically heated

Iron by Crompton Limited, 1891.

pad. This proved fairly inefficient since they lost their heat quickly and the search was soon on for an iron directly heated by electric power. By the early 1980s, both GEC and Crompton were selling such models in England. The early electric models exactly resembled their pre-electric ancestors – a heavy iron base with straight sides, a slight point at the front and a wooden handle. The electric socket was positioned at the back and the plug attached to the light socket. This remained the case for a couple of decades and numerous illustrations in books about household work in the early days of the century exist depicting women ironing with their irons attached to overhead sockets.

In spite of its availability the electric iron didn't become a popular household appliance in the USA and Europe until the 1920s and 1930s. In the USA, however, the Hotpoint Company, so named because it sold an iron with as much heat in its tip as in its middle, was selling considerable numbers of irons as early as 1905 and in 1912 the American Heating Company introduced its 'American Beauty' model which was the first commercial iron aimed directly at the household. Like washing-machines, the electric iron had its best reception in commercial laundries where efficiency and speed were essential and it was from there that if found its way into the domestic environment. The 'American Beauty' was also the first modern-looking iron. With its curved, steel sides it pioneered the 'streamlined' look, widespread by the middle of the century. Electricity was not the only energy source for ironing at this early stage, however, as gas, petrol, and alcohol-powered models also played their parts in speeding up the ironing process.

Electric irons were, inevitably, a more costly investment than their alternatively powered counterparts and, with the servant problem being less acute in Great Britain than in the USA it took longer for British housewives to realize the benefits of electric ironing. In the USA, however, the 1920s saw an enthusiastic reception for the domestic electric iron and it became one of the most popular household appliances. Industrial designers made it look increasingly clean, efficient and appealing by chroming the body and integrating the, by now, bakelite handle more carefully into the overall form of the object.

The next significant breakthrough also occurred on American soil. It was in 1927 that the first adjustable automatic thermostat was introduced into the electric iron. The resultant 'controlled heat' iron was a vast improvement upon earlier models as now not only did irons not overheat and burn clothing they could also have the degree of their heat modified according to the kind of fabric that was being used. With the advent of synthetics as popular alternatives to natural fabrics in the 1930s, lower ironing heats were needed and the addition of the thermostat and adjustable heat settings made this possible for the first time. Inevitably, though, 'automatic' irons were more expensive than their non-automatic counterparts and the early models were also somewhat unreliable. It was not, in fact, until the end of the 1930s in the USA that the inclusion of an automatic iron in the household inventory had become widespread.

A popular alternative to the electric iron in the USA in the 1930s was the 'electric ironer', a machine described at the time as being 'a unique example of large-scale commercial facilities now made possible

Above: 'Mary Ann' iron from the Thorn Electrical Company, 1946.

Left: Rowenta steam-iron DA21, 1980.

for home application'. Whether of the 'rotary' type or the 'flat presser' variety the 'ironer', complete with a bench to sit on, was seen by many as a more efficient and less arduous means of ironing large pieces of fabric such as sheets and table-cloths. Because of its bulk, however, it gradually disappeared during the 1930s and was rarely seen in the home after the Second World War.

In Great Britain in the 1930s and 1940s, a number of advances were made in iron design as it began to become an increasingly popular household item in that country as well. Christian Barman's little streamlined, controlled heat iron for HMV, covered with primrose yellow porcelain enamel, was a particularly advanced model combining a sophisticated ergonomic approach with a new bright clean appearance. Through the 1940s a number of British companies came up with irons combining thermostatic controls with coloured enamelled bases and curved black bakelite handles. While still relatively heavy by today's standards, these models were considerably lighter to handle than the old flat irons and they could be used quite easily by the housewife. Novel examples included Thorn Electrical Company's 'Mary Ann' iron, which had

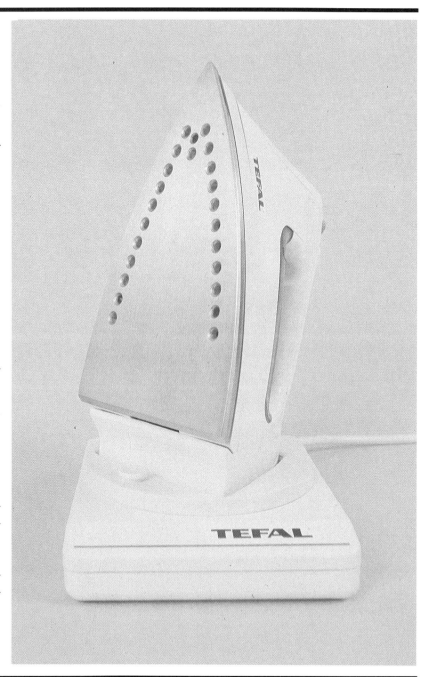

Tefal cordless iron, 1980s.

points at both ends to facilitate ironing areas awkward to penetrate and the designer Sadie Speight proposed a model (sadly never produced) for the Beethoven Electrical Equipment Company, which had a ridge at its base to enable ironing beneath buttons. The Morphy Richards Company captured a large share of the post-war British market with its 'PA75' model, and went on, through the 1960s, to produce some of Britain's most popular irons.

It was in the USA, however, that all the major technological advances were made and this was equally true of the next innovation in iron technology – the steam iron. The first American steam iron, the 'Eldec', was available on the market as early as 1926 but it was to be some time before it achieved widespread popularity. By the early post-Second World War years, however, most American homes possessed either a steam iron or a combination steam/dry iron. By 1950, for instance, Universal was selling its 'Stroke-Sav'r', a highly streamlined iron which emitted 'vapour steam' and didn't need to have distilled water put into it like so many earlier models. Steam irons became increasingly popular in the USA through the 1950s and, in Britain and Europe through the 1960s and 1970s. As housewives became the exclusive users of irons the latter rejected their old 'black and sooty' image and became bright appendages of the modern electrified home.

With heat control and steam the idea that heavier irons did the job more efficiently than light ones became redundant, and, with the inclusion of more and more plastic components, the post-war iron became a completely new object. The distinction between the body and the handle became increasingly eroded and gradually the whole body and handle became a single plastic unit while the base remained steel. By the 1980s, irons could be bought in a range of colours to match the interior decor.

In addition to the standard household iron, the travelling iron has continued to fascinate both manufacturers and users since the very early days of ironing. Inevitably travelling makes ironing a necessity (although less so since the advent of crease-resistant fabrics) and many small, foldaway irons have been produced over

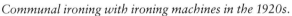

Communal ironing with ironing machines in the 1920s.

Right: Demonstration of iron which cannot scorch materials even if left in position. British Industries Fair, 1952.

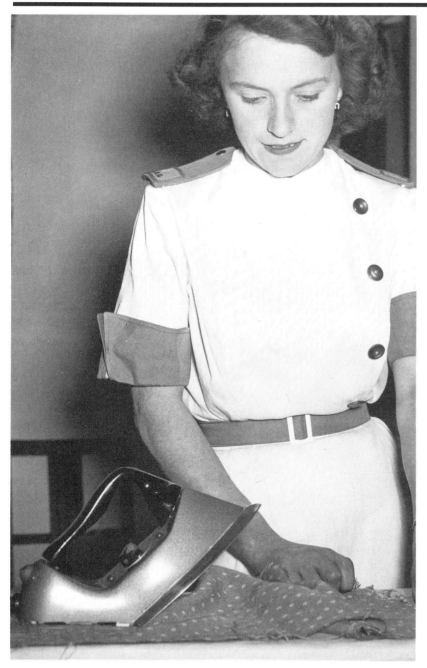

the years to meet that particular need. In Britain the little 'Clem' iron, for instance, was a popular object in the 1930s and 1940s while more recently irons which fold completely flat have been put on the market by a number of manufacturers, Japanese ones among them.

Over the last few decades, with the increased availability of synthetic fabrics, people have continually predicted the end of ironing. It seems, however, that, whatever the appeal of artificial textiles, natural fabrics, such as cotton, continue to find a place on the market and, as a result, ironing remains a fairly common domestic task. It seems unlikely that, in the foreseeable future at least, the domestic electric iron will disappear from modern life. Ironically, one of today's latest proposals where ironing is concerned takes the story of the modern electric full circle. Designers and manufacturers have, since the 1880s, been in continual search for a cordless iron to eliminate the need for a flex suspended from the iron to the electric socket. The most recent example on the market is not a new idea but rather an old one which has been improved. It works on the basis of the iron being recharged every time it is replaced at the end of the ironing board during use. It remains to be seen whether this idea will provide the basis of the next major breakthrough in modern iron design.

8

CLEANING AND HEATING THE HOME

"Earlier and more conspicuously successful than all other mechanised household appliances, the vacuum cleaner, which can be put away in any broom closet, made its way through the world."

S. Giedion 1948

AFTER irons, vacuum cleaners were the best selling domestic appliances of the first half of this century. They represented a real means of saving existing labour and they were seen as a necessity to health at a time when it was generally thought that germs gathered in household dust. The vacuum cleaner was seen as a blessing in numerous households and it undoubtedly speeded the process of cleaning the home.

The term 'vacuum' is, however, a misnomer as the cleaner actually works on the principle of suction – it literally sucks dust and dirt out of rugs and carpets. This was most commonly done through the use of a motor-driven fan which caused dirt to be pulled in through the nozzle.

The invention of the 'vacuum cleaner' took place in Great Britain. H. C. Booth was responsible for inventing the principle and in 1901, he went on to form the Vacuum Cleaner Company Limited, changing its name, in 1926, to the better known 'Goblin (BVC) Limited'. As usual, however, it was in the USA that the first lightweight, portable, domestic suction cleaner was developed. In the 1890s, the USA had been at the forefront of developing the stationary model of suction cleaning which was applied fairly widely in commercial buildings at the time. This continued into the 1920s but had little influence on the domestic sector which became dominated by the portable models.

The patent for electrically-driven suction cleaning was given to the American David Kennedy in 1907 but the idea of the portable 'bag and stick' cleaner was conceived in the same year by another American called Murray Spangler. Although the latter patented his idea in the following year and set up his own business – the Electric Suction Sweeper Company – his poor financial situation led him to be bought by the Hoover Company in the same year. The first cleaner, the 'Model O', manufactured by Hoover was, as we have seen, a very simple 'bag and stick' model with a crude metal casing covering the brush, fan and motor. Aggressive sales and advertising succeeded in persuading American housewives that this new product would help in their housework and reduce labour.

By 1910 it had become obvious that there was a receptive market for the vacuum cleaner in the USA and a number of other manufacturers, among them Eureka and Regina, provided alternative models to Hoover's. General Electric introduced a portable 'drum' model in 1914, a form which became popular once again after the Second World War.

In Britain, in the early century, vacuum cleaners were much more bulky, consisting as they did of a set of components mounted on a trolley, and they were restricted, for the most part, to the commercial sector.

Although the Hoover portable model was available in Britain as early as 1912 it wasn't until 1921 that Booth introduced a small domestic model to compete with it. Through the 1920s a number of upright cleaners were available in the USA, Britain and Germany, all of them based upon the same format – that of a cast metal housing for the mechanical components combined with a black bag to hold the dust and a broom-stick handle. When, in 1922, an American commentator remarked that 'all vacuum cleaners look charming and shiny' he was referring, undoubtedly, to the effect of the metal body-shells.

It was in the 1920s also that the first real competitor to the upright model emerged. In 1919 the Swedish company, Electrolux, produced a 'cannister-type' vacuum cleaner which glided across the floor on two metal skis. The advantages of this model were its numerous attachments which could be fixed on to the hose and, as a result, the cleaner could be used to dust curtains, bookshelves, stairs and a number of other previously untouched areas in the home. Electrolux brought its cannister cleaner to the USA in 1924 and to Britain a couple of years later and it very quickly became one of the most

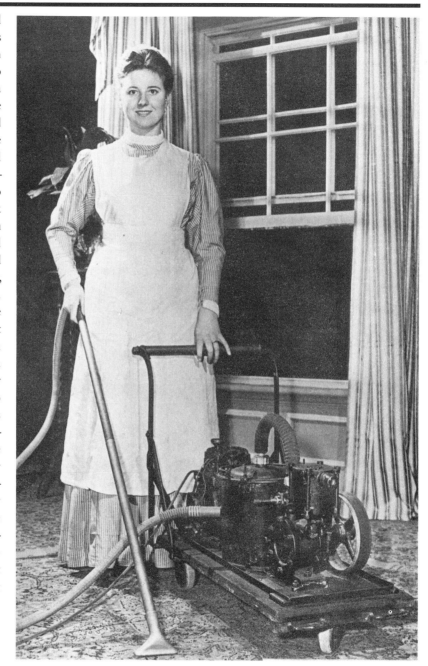

Goblin vacuum cleaner of the 1900s.

'Wizard' vacuum cleaner, 1911.

popular models on the market. Like the upright Hoover cleaners, the Electrolux models were sold by salesmen on a house to house basis rather than through retailers and, in the inter-war years, the concept of the 'vacuum-cleaner' salesman became widespread in a number of countries.

By the 1930s, a number of other American manufacturing companies joined the existing ones in making and selling vacuum cleaners. Montgomery Ward and Sears Roebuck produced their own ranges (the former designed by W. D. Teague) and both Sunbeam and Singer also joined the competition. The 'upright' and 'cannister' models were the most popular and it was in this decade that the industrial designer took on the task of making them look less like a combination of components and more like modern streamlined domestic artefacts. Henry Dreyfuss' design for Hoover of 1936, the 'Model 150' stood out as one of the most successful attempts to create a character for the modern vacuum cleaner – it had a more visually unified base and he updated the typography on the dust-bag. Other innovations were the use of magnesium, making it lighter and its plastic hood. Many other models emulated it in the years that followed. Novelty features, such as a light to illuminate the part of the floor that was being cleaned, and a small hand-held suction cleaner, used for such jobs as cleaning furniture and the insides of automobiles, were also introduced in this decade.

Another big leap in vacuum cleaner ownership occurred in the USA immediately after the Second World War. While, for instance, approximately one million cleaners had been bought in 1920, by 1947 this figure had risen to fourteen million. A new development in the 1950s was the re-emergence of the 'tub' model, a hybrid cleaner half-way between an upright and a cannister model. It had all the advantages of the latter as numerous attachments could be fitted to it, but it was not so close to the ground and could there-fore be switched on and off with more ease. In 1950, for instance, Lewyt advertised such a cleaner pro-claiming that there was 'no dust bag to empty' and General Electric presented a similar model with a 'swivel-top for "reach-easy" cleaning'.

Smaller, more powerful motors were developed in the post-war years and, as a result, vacuum cleaners became more efficient and easier objects to move around the house. A great emphasis was laid upon the fact that, with little wheels, the 'tub' mod-els could be pulled around the house behind the user. Electrolux' models lost their metal gliders in this period

and in 1955 Hoover's new spherical 'Constellation' cleaner had a swivel top and an 'easy-glide' base. This remarkable cleaner came complete with a stretchable hose which reached all sorts of nooks and cran-nies. Visually the 'Constellation' rep-resented a move, in the aesthetic of domestic products, away from automobile styling and towards the space-age imagery which was to dominate the second half of the 1950s in the USA. Its plastic body shell came in two shades of blue – a shift away from the ubiquitous black and chrome of the upright model and introduced the possibility of integra-tion into the interior decor of the home. So evocative was the imagery of the 'Constellation' that the English artist Richard Hamilton used it in his famous 'Pop' collage, 'Just what is it that makes today's home so different, so appealing' of 1956.

Colour remained a characteristic of vacuum cleaner design through the 1950s. In Italy the Castiglioni brothers' little 'Spalter' cleaner had a bright red plastic case while other American models, such as Cadillac's 'Just-Vac' cleaner, boasted plastic shells in a variety of pastel shades. In 1959 over 70 per cent of American households owned a vacuum cleaner.

In Britain, however, things had not progressed so quickly and the models available after the War were still based on the 'bag and stick' and

'cannister' types. However, the Elec-trolux had lost its gliders and squared off its sides and the upright Hoover was a little simpler in design than its pre-war equivalent.

From the 1960s onwards the use of plastics as the main material used in vacuum cleaner bodies expanded enormously. As a result, ranges of machines with highly compact forms emerged. In the upright model the exposed fabric bag was replaced by a plastic body containing a dust-bag — cleaners began to look less and less like sets of components. European manufacturers, among them Miele, Philips and Rowenta, entered the competition in these years along with AEG which had moved on a long way from its formidable 'Vampyr' model of the 1920s. In the 1970s and 1980s, marketing managers introduced such evocative and persuasive terms as 'turbomatic' in an attempt to bring an impression of advanced technology into the world of vacuum cleaners.

One fairly revolutionary cleaner from the 1980s was the 'Cyclon' model, designed by the Englishman James Dyson but manufactured by a Japanese company. In Post-Modern colours, pink and lilac, recalling those of the 1950s – it had a radically new appearance and also introduced a new method of sucking in dirt using cyclone technology which signific-antly increased the sucking power of the machine.

AEG's 'Vampyr' vacuum cleaner, 1921.

Above: Hoover Limited: 105 'Senior' vacuum cleaner, 1920.

Right: Panasonic vacuum cleaners of 1986.

Below: James Dyson's prototype for his 'Cyclon' vacuum cleaner, 1980s.

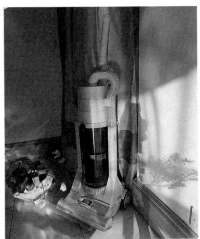

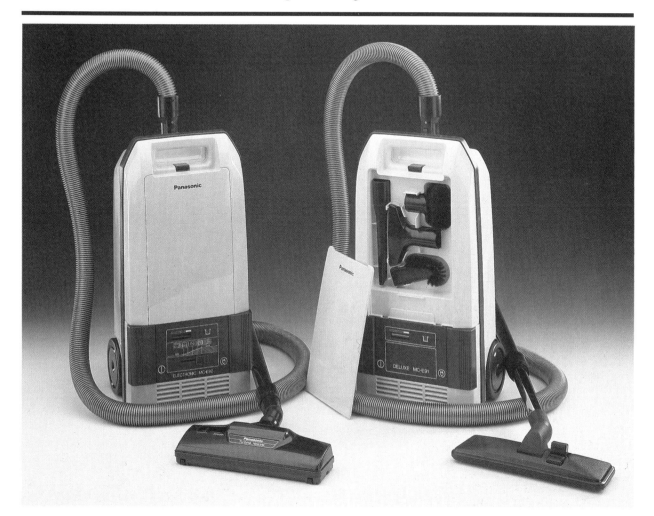

Allied to the vacuum cleaner is another product which has all but died out in the domestic sector. The electric floor polisher was a popular item from the 1920s onwards. Regina produced its first model in 1929, and with the fashion for parquet floors and linoleum from the 1930s through to the 1950s the floor polisher came into its own. With the advent of fitted carpets, however, it is now a rare sight in the home.

The vacuum cleaner has, in contrast, continued to maintain a strong presence in the domestic environment and has remained one of the most popular and widely used of today's appliances. Three models, the upright, cylinder and tub (often used in a commercial context) still vie with each other, each one providing features which are exclusive to it alone and which fit it for a particular context. It seems unlikely, at least in the foreseeable future, that the vacuum cleaner will cease to be a familiar feature of contemporary domestic life.

"AROUND 1945 the electric dishwasher in a sense still belonged among shelved inventions. Its production for the mass market will doubtless have to await such time as the servantless household becomes universal."

S. Giedion 1948

In her treatise of 1913, *Scientific Management in the Home*, the American writer Christine Frederick included an illustration depicting a woman doing 'efficient' washing-up. It showed her sitting on a high stool with, to her right, a trolley on wheels which had all the dirty dishes stacked up on it and, to her left, open shelving where she placed them as soon as they were clean. What Mrs Frederick could have been looking at, if she was really concerned with saving time in washing-up, was the new electric dishwasher emerging on to the market at that time, although it was still rather primitive and very expensive. By 1917, in her book of that year, Mrs Peel illustrated a 'Dreadnought' washing-up machine which, she wrote, 'may be worked by hand or electricity'. Crudely made and lacking visual integration it simply consisted of a metal box with a handle which rotated a drum. The dirty utensils were held inside a metal cage and the whole assemblage sat on a four legged plinth.

Five years later in the USA a 'portable' dishwasher was discussed. Equipped with a motor it contained a 'fan' or 'paddle' which whirled the water among the dishes. With the housewife doing so much of her own work by this time it had become clear that the electric dish-washer was potentially a very useful utensil. While the portable model was essentially a drum on wheels, a larger rectangular model with four legs was also available. When not washing dishes it could be used as a table. A decade later the automatic electric dish-washer could be combined with the sink and concealed behind one of the doors of the sink units. A writer from the 1930s explained that, "*the dish-washer will wash dishes cleaner, more hygienically and more quickly than human hands, because hotter water can be used.*" By the outbreak of the Second World War, the dishwasher had become a fairly established item in the American kitchen along with the refrigerator and the electric range. There was some resistance to it – Ethel R. Peyser maintained, for instance, that "*I think a dishwasher is a nuisance, you have to stack your dishes, hand-scrape pots and pans, cart water by the pailful and then you have the job of cleaning the dishwasher itself.*" But the benefits of the appliance were also beginnning to be appreciated, although it was seen as more of a luxury than a necessity. It took quite some time to become very popular.

It was in the post-Second World War years that the dishwasher finally became a ubiquitous object in the USA, now not only in affluent urban households but in less wealthy rural and suburban ones as well. Along with the freezer, the steam iron and the waste disposal unit, the dishwasher was widely manufactured by, among others, General Electric, Hotpoint, Kitchenaid, Westinghouse and Frigidaire – and sold after 1945. Kitchenaid's model from 1955, for instance, was a square, front-loading washer, available with a white enamel, stainless steel or copper finish. It fitted neatly under the kitchen surface alongside the other major appliances. General Electric had produced the first square tub well before the War but it wasn't until the mid-1950s that this design was featured in popular women's magazines.

If the USA was, at first, fairly reticent about the advantages of the electric dishwasher, Great Britain was, predictably, even more so. A *Good Housekeeping* Encyclopaedia of the mid-1950s claimed, for instance, that while it *was* worthwhile for a household of 12 or more to own a dishwasher, for the smaller family it was considered a waste of time. As the writer maintained, "*the time taken in loading, unloading and cleaning the machine is often more than that saved in carrying out the actual washing-up process.*" Even in 1962 a *Design* magazine report on Prestcold's 'DP 101' machine, modelled on the American company RCA's 'Whirlpool' design of a few years earlier, laid more emphasis on the things it couldn't do than on those it could.

The main reason for this resistance was a combination of general conservatism and the fact that most of the machines available on the British market were top-loading and had to be wheeled across the room to be attached to the water-supply. This was, inevitably, fairly time-consuming and cumbersome, and most households couldn't justify the expense either. In the same 1962 article, Bruce Archer went as far as to claim that the electric dish-washer was "*an answer to marketing problems (inasmuch as it allowed companies such as Pressed Steel to use up their sheet steel supplies) but not to basic use problems.*"

Eventually, with improvements made to the automatic dishwashers available on the British market matching those in the USA, the British public began to see, or were encouraged to see, the advantages of the machines, particularly as they began to decrease in cost. Britain still has some way to go, however, before it matches the Americans and Euro-peans in its wholehearted commitment to the electric dishwasher.

"*ELECTRIC heating may be as successful at some future date as electric lighting itself*"

A. E. Kennelly 1890

Although heating by electricity has always been more costly than its equivalent by solid fuel or gas it has, nonetheless, been in constant use since the turn of the century. Indeed, electric heating was an inevitable product of electric lighting. As A. E. Kennelly explained in 1890 "*the*

GEC 'sausage' bulb heater, 1906.

problem and task of electric lighting is, primarily and essentially, electric heating", and it was not long before inventors had found ways of tapping that source, both, as we have seen, for cooking purposes and, simply, for heating the house.

The earliest form of electric heating was with what were called 'lamp radiators'. Invented in 1896 by an Englishman called Dowsing, these fires had large 'sausage' bulbs which gave off heat through radiation: Cannon's fire of 1904, for instance, consisted of four of these 'sausages'. This method was rapidly superceded, however, by two other forms of electric heating which remained the norm for electric fire technology until well into the second half of this century. In 1910 Dr. S. Z. de Ferranti discovered the principle of the parabolic reflector electric fire and two years later Belling, an ex-Crompton employee, found that he could create heat by winding resistance wire around a strip of fire-clay. He patented his discovery in 1912 and, in the same year, launched his 'standard' fire which remained in production until 1927. It consisted simply, of a metal frame, a fender, two switches, a ruby lamp and six fire bars. As the electric fire was a living-room, or bedroom, item rather than a kitchen utensil, Belling's model was appropriately decorated with Art Nouveau-inspired patterns on its metal surround, and, in imitation of the coal fire that it replaced, a trivet for the kettle was also provided.

The idea of the electric fire which stood in the grate, providing a clean and easy, if expensive, alternative to an open fire emerged, therefore, in Britain in the years before the First World War and it remained a popular household item for a number of decades. In 1917, for instance, in her book *The Labour-Saving Home*, Mrs Peel illustrated a 'Baby' reflector fire, made by the British Electric Transformer Company which could be moved into a horizontal position to accommodate both a kettle and an iron.

In the USA, in sharp contrast, the preference was for 'air' or 'space' heaters which didn't provide a red glow or act as a direct substitute for an open fire. These were large metal cabinets, with grid work on their surfaces. They caused one writer to comment, in 1914, that *"The systems of hot air heating... invented over the last fifty years, are great improvements over the old methods of heat-*

HMV convector heater designed by Christian Barman in 1934.

ing houses. They have however, destroyed the social influence of the fire." This tendency remained the norm, in the USA until the 1930s by which time two types of space heater were available – the 'natural draft' heater, consisting of a row of vertical copper ducts set out into a frame, and the 'forced draft' version through which the air was blown by a fan. General Electric's model had a streamlined metal frame surrounding it with a circular grid on the front. The 'duct' version was, by comparison, much more old-fashioned in appearance – a rectangular box with four little legs and a shelf on the top. Radian heaters, both with and with- out automatic thermostats, of the bowl type and others, as well as 'electric grate' heaters were also available in the USA at this time. They were outnumbered by the space heater though, which in turn was gradually replaced by the radiator in the majority of American homes in the inter-war years.

Radiators, which could heat the whole house, rather than merely one

HMV 'Cavendish' convector heater, 1946.

part of it a time, were a logical choice for the USA which, on the whole, had harder winters than in Great Britain. By the 1930s they could either be hung on the wall or were set in. They became standard equipment in most of the new urban houses and apartments built in that decade.

In Britain the 'grate' or 'portable' electric fire remained popular through the 1920s and 1930s. Back in 1913, the Falkirk Iron Company had manufactured the first imitation coal fire and in 1920 H. H. Berry improved on it by adding a flicker effect, giving the impression of an authentic coal fire. This style of fire became very popular in Britain as the electric fire continued to provide an important visual and psychological focus in the living-room and a number of companies produced their own versions. In the mid-1930s, for instance, as well as manufacturing pedestal bowl fires, reflector fires, inset fires and lamp radiators GEC produced a range of 'Magicoal' fires in a number of different styles with such evocative names as 'Wonderberry', 'Sundown' and 'Fireblaze'. While some fires were in the modern 'Art Deco' style, most were period revival pieces – unlike their kitchen counterparts, electric appliances for the living room tended to be designed in traditional rather than modern styles.

There were a few notable exceptions to this rule, however. A number of British fires of the 1930s and 1940s were the results of companies bringing in consultant designers to provide them with innovative designs. The work of Christian Barman for HMV and Wells Coates for E. K. Cole were amongst the most notable. Barman, for instance, designed a portable radiant fire in aluminium which sat in a spherical basket guard and for HMV he created a convector heater consisting of a series of horizontal semi-circular bands of chromed steel – a highly dramatic design. It was modified a little later and in 1946 re-emerged as HMV's 'Cavendish' convector heater. This particular form of heating provided the main alternative to the reflector fire in the inter-war

Right: Morphy Richards Limited. Portable electric radiant fire, 1954.

Left: Belling Limited. Imitation coal fire manufactured by Belling Limited between 1921 and 1927.

and immediate post-war years in Great Britain. Other innovative British portable electric fires of these years included designs for Ferranti, created by their resident artist W. N. Duffy. Novelty designs were also immensely popular, among them chromium-plated electric bar fires in the form of sailing yachts. Many either still fitted into the grate or sat in front of the fire-place, providing a visual focus for the room. With the advent of television into most homes, however, the electric fire became less crucial in this capacity and as a result, their designs became less and less extravagant.

It wasn't until the mid-1950s that the electric fan heater became available on the British market. The USA had used the same principle in its space heaters back in the 1930s but it was in Germany that the concept was refined and presented in the form of a small, floor-based fan-assisted heater. It remains, today, one of the most convenient and rapid ways of heating a single room.

Throughout its history, electric heating has competed with other power sources and gas has consistently provided a very competitive alternative, both for small fires and for central heating systems. Electricity remains, however, used fairly widely for domestic heating although the portable fire is less ubiquitous than it was 20 years ago. Central heating is now the dominant form of domestic heating in the USA, Britain and Europe and in many ways the individual electric fire has become increasingly a thing of the past.

Hoover Limited. Fan heater 3000, 1966.

CONCLUSION

THE 'MYTH' OF LABOUR - SAVING DEVICES

IN the second half of this book I have charted some of the key developments in a number of the most significant of this century's domestic appliances. While they represent many of the crucial stages in the formulation of today's appliances, that evolution has not taken place according to some abstract law of 'progress' but rather as part of the more general socio-economic and cultural movements of this century.

Thus, as I have already demonstrated, it is impossible to isolate the evolution of contemporary domestic appliances from, among other things, the changing status of domestic servants and the evolving rôle of women in the home. It is also important to take into account the parallel developments that have taken place within modern technology, industry and commerce. Without, for example the dominance of the large-scale American manufacturing companies which moved into appliance production, the evolution of modern domestic appliances might have taken a number of diffe-rent routes. Also, without the compulsion to produce new models to reach ever new markets, the industrial designer would have had a much lower profile.

As we have seen, the question of design, and of appliance design in particular, cannot be separated from its complex context. The designer operates as part of a system beginning at production and ending with consumption and use. Uniquely, his link in that chain is to actually propose the forms that will eventually transmit the messages to the consumer. Whether named or anonymous, alone or a member of a team, the designer in the appliance industry has communicated the 'myths' which have caused electrical appliances to sustain their central rôle within every day life. They have also played an important part in dictating the rôle that women should play in the home, creating, through their products, the mythical, idealized personae which were to accompany them. Frequently their idealizations have had a strong, influence upon the world of reality.

Inevitably, through this century the myths have evolved and changed. At the turn of the century, for instance, the rôle of tradition was important in persuading the new consumer of, say, an electric coffee-pot. This new-fangled object needed to be seen to be as valid a household utensil as the old one which went on the paraffin heater. Electric coffee-pots therefore emulated, with the help of the designer, their precursors. It wasn't until the 1920s, when electricity as a source of heat and power had become more generally accepted, that manufacturers and designers were able to move beyond the concepts of 'tradition' and sheer 'utility' as selling tools for electrical domestic appliances. Now they began to promote a new, rational, aggressively modern, aesthetic for their products implying progress and, more specifically, freedom from household chores for the modern woman.

At a time when servants were less in evidence and the rôle of the housewife was becoming more universal, appliance designers and advertisers promoted their goods as the answer to women's problems and the providers of leisure. It was at this time that the concept of the 'labour-saving' device was invented and became deeply embedded in the public's consciousness through intensive advertising campaigns.

When the cause of female liberation became a little overworked in the inter-war years it was joined by another powerful sales strategy – the link between electrical appliances and hygiene. Again it was the designer who created the forms which transmitted the myth and the gleaming white products of the inter-war years bear witness to his achievements. Ironically, however, as a number of writers on the subject have pointed out, this myth cut across the earlier one of female emancipation as it brought with it the requirement for higher standards of hygiene and cleanliness in the home.

In the years following the Second World War, appliance design was at the peak of its demand. The designer came into his own as the creator of 'jewellery' for the kitchen – of precious objects which had as much significance as status objects as items of utility. The consumers of these years understood design as being the use of expressive form and colour which had become the *sine qua non* of electrical goods.

The housewife found herself once more trapped in the kitchen, now no longer simply as the user of these appliances but as the 'glamorous hostess' proud to show her guests her new acquisitions which prepared their meal, as if by magic, with the minimal amount of assistance from her.

After the 1950s, the appliance designers retreated temporarily from a world of fantasy to one in which efficiency and modernity dominated the picture once again. The housewife, busier than ever before with the rising graph of female employment in these years, could only be won over by the promise of labour saved. In many cases the promise was a false one. Much of their newly acquired earnings went on consuming the very products which, theoretically, allowed them to earn in the first place.

The final social myth perpetrated by electrical appliance designers and manufacturers to date recalls that of the early century once again. Disillusioned by the constant appeal to

Prototypes of 'Post-Modern' designs by National Panasonic, 1986.

modernity and the promise of progress, many appliances, together with a number of other household items, have reverted to an image of the past – to that of a rural, pre-industrial idyll untouched by the horrors of industrialization and mechanization. Once again it is the designer who has found ways of expressing this longing, this time through his use of earthy colours and rural imagery. The sprigs of corn and poppies adorning the sides of toasters and kettles are no arbitrary motifs, but part of the mood of 'retrospective regret' that dominated the 1970s and early 1980s.

In creating the myths and images around us, it must be realized that the designer has not created a set of cultural norms, but merely reflected and sustained them. For this reason appliance design in this century has supported the cultural *status quo* and, indeed, depended upon it for its mass consumption and acceptance in the home. It is for this reason that, in wanting to attack both the patriarchal and capitalistic norms of our times, so many writers have focused on domestic appliances as not being merely the symbols but the actual perpetrators of the values of the *status quo*. They blame, for instance, complex domestic machines for creating rather than saving labour; individualized machines for breaking up the community; household technology for alienating the housewife from what was previously fulfilling work in the home; and the desire for the consumption of domestic appliances as part of the reason for women entering into paid, rather than domestic, servitude. In many ways they are, of course, right. Electrical appliances have been produced by the sharp end of capitalistic manufacturing industry and they *do* represent all the social norms that these writers wish to oppose.

At the same time, however, dishwashers, *do* now save time and labour, vacuum cleaners *do* make cleaning an altogether easier task, and freezers *do* minimize shopping trips. Ignoring for a moment the ideological framework within which domestic appliances are both made and used, the potential for many of them to improve our lives is ever present. A passage from Iris Murdoch's novel *The Good Apprentice*, for instance, describes a house in which old and new appliances are combined, the former for their continued usefulness and their associations with a better 'past', and the latter for their use value: "*There was a large handsome kitchen, with a long cast-iron cooking stove... the wash room with a washing machine and trapeze-like wooden drying frame, and the 'brushing room' full of dustpans and brooms and boots and shoes, which also housed the enormous deep freeze.*"

The fictional owner of this kitchen had refuted the idea of progress for progress's sake, refused to be sold appliances through their advertisements and has selected, instead, those appliances from the past and present which continue to adequately serve their intended function. The time saved by using them was, in the novel, then utilized for more creative household pursuits such as pottery and carpentry.

It is only with this kind of discerning attitude towards the appliances with which we surround ourselves that it might be possible to fulfil the ever hopeful promise that 'labour saved *is* leisure gained'.

BIBLIOGRAPHY

ADAMSON, G. *Machines at Home* Lutterworth Press, London 1969.

BEECHER, C.E. *A Treatise on Domestic Economy for the Use of Young Ladies at Home and at School* New York 1845 (Third Edition).

BEECHER, C.E. and BEECHER STOWE, H. *The American Woman's Home* New York 1869.

BINNIE, R. and BOXALL, J.E. *Housecraft: Principles and Practice* Sir Isaac Pitmans and Sons, London 1926.

BRIGHT, A.A. *The Electric Lamp Industry: Technological Change and Economic Development from 1800–1947* Macmillan, New York 1949.

BROOKE, S. *Hearth and Home: A Short History of Domestic Equipment* Mills and Boon, London 1973.

BURCKHARDT, F. and FRANKSEN, I. *Design: Dieter Rams* Gerhardt Verlag, Berlin 1980.

BURCKHARDT, L. (ed) *The Werkbund* Design Council, London 1980.

BYATT, I.C.R. *The British Electrical Industry 1875–1914* Clarendon Press, Oxford 1979.

BYERS, A. *Centenary of Service: A History of Electricity in the Home* Electricity Council, London 1981.

BYERS, A. *Home Electricity* Pelham Books, London 1969.

BYERS, A. *The Philips Key to Electric Living* Philips, London 1975.

CORLEY, T.A.B. *Domestic Electrical Appliances* Jonathan Cape, London 1966.

COWAN, R.S. *More Work for Mother: The Ironies of Household Technology from the Open Hearth to the Microwave* Basic Books, New York 1983.

DAVIDSON, C. *A Woman's Work is Never Done: A History of Housework in the British Isles 1650–1950* Chatto and Windus, London 1982.

De HAAN, D. *Antique Household Gadgets and Appliances 1860–1930* Blandford, Poole 1977.

DUDDEN, F.A. *Serving Women: Household Service in Nineteenth-Century America* Wesleyan University Press, New York 1983.

EVELEIGH, D.J. *Fire Grates and Kitchen Ranges* Shire Publications, Aylesbury 1983.

FALES, W. *What's New in Home Decorating?* Dodd Mead and Co., New York 1936.

FAULKNER, W. and ARNOLD, E. (eds) *Smothered by Invention* Pluto Press, London 1985.

FAUNTHORPE, REV. J.P. *Household Science: Readings in Necessary Knowledge for Girls and Young Women* Edward Stanford, London 1895.

FEARN, J. *Domestic Bygones* Shire Publications, Aylesbury 1977.

FINN, B.S. (ed) *Lighting a Revolution* Smithsonian Institution, Washington 1979.

FREDERICK, C. *Household Engineering* American School of

Home Economics, Chicago 1913.

FREDERICK, C. *Selling Mrs. Consumer* New York 1929.

FREGNANT, D. *Electrical Collectibles: Relics of the Electrical Age* Padre Publications, California 1981.

GIEDION, S. *Mechanization Takes Command* W. W. Norton and Co., New York 1969.

GILBRETH, L. *The Home-Maker and Her Job* D. Appleton, New York 1927.

GLOAG, J. *Industrial Art Explained* George Allen and Unwin Ltd, London 1946.

GLOAG, J. *The Missing Technician* George Allen and Unwin Ltd, London 1944.

GLOAG, J. *Plastics and Industrial Design* George Allen and Unwin Ltd, London 1945.

GORB, P. (ed) *Living by Design* Lund Humphries, London 1979.

GORDON, B. *Early Electrical Appliances* Shire Publications, Aylesbury 1984.

GOWANS WHYTE, A. *Forty Years of Electrical Progress: The Story of GEC* London 1930.

GOWANS WHYTE, A. *The All-Electric Age* London 1922.

GREEN, H. *The Light of the Home: An Intimate View of the Lives of Women in Victorian America* Pantheon Books, New York 1983.

GRIES, J.M. and FORD, J.

(eds) *Household Management and Kitchens* National Capital Press, Washington 1932.

HARLAND, M. *House and Home: A Complete Housewife's Guide* Philadelphia 1889.

HARRISON, M. *Home Inventions* Usborne Publications, London 1975.

HARRISON, M. *The Kitchen in History* Osprey, Reading 1972.

HAYDEN, D. *The Grand Domestic Revolution* The MIT Press, Cambridge Mass 1981.

HENNESSEY, R.A.S. *The Electric Revolution* Oriel Press, Newcastle 1972.

HOBBS, E.W. *Domestic Electric Appliances* London 1930.

HOLMES, G. *Industrial Design and the Future* London 1934.

HOLT, J.M. *Housecraft Science* G. Bell & Sons Ltd, London 1953.

JONES, R. and MARRIOTT, O. *Anatomy of a Merger: A History of GEC, AEI and English Electric* London 1970.

KATZMAN, D.M. *Seven Days a Week: Women and Domestic Service in Industrializing America* University of Illinois Press, Chicago 1981.

LANCASTER, M. *Electric Cooking, Heating, Cleaning etc.* Constable, London 1914.

LINCOLN, E.S. and SMITH, P.C. *The Electric Home* Electric Home Publishing Co., New York

1934.

LYNDE, C.J. *Physics of the Household* Macmillan, New York 1914.

MACLAREN, M. *The Rise of the Electrical Industry in the USA* Princeton, New York 1943.

McBRIDE, T.M. *The Domestic Revolution: The Modernization of Household Service in England and France 1820–1920* New York 1976.

MARCHAND, R. *Advertising the American Dream* University of California Press, Los Angeles 1985.

MERCER, F.A. *The Industrial Design Consultant* The Studio, London 1947.

NOVY, P. *Housework without Tears* Pilot Press, London 1945.

OAKLEY, A. *Housewife* Penguin, Harmondsworth 1976.

PASSER, H.C. *The Electrical Manufacturers 1875–1900* Harvard University Press 1953.

PEEL, Mrs C.S. *The Labour-Saving House* London 1917.

PEET, L.J. and THYE, L.E. *Household Equipment* Chapman and Hall, London 1940.

PEYSER, E.R. *Cheating the Junk Pile* Dutton and Co., New York 1922.

PHILLIPS, R.R. *The Servantless House* Country Life, London

1923.

RANDELL, W.L. *Electricity and Women: 21 Years of Progress* Electrical Association for Women, London 1945.

SAMBROOK, P. *Laundry Bygones* Shires Publications, Aylesbury 1983.

SIMPSON, H. *The Happy Housewife* Hodder and Stoughton, London 1934.

SKLAR, K.K. *Catherine Beecher: A Study in American Domesticity* New Haven 1973.

STRASSER, S. *Never Done: A History of American Housework* Pantheon Books, New York 1982.

TERRILL, B.M. *Household Management* Chicago 1914.

THOMPSON, H. *The Age of Invention* New Haven 1921.

VAN DOREN, H. *Industrial Design: A Practical Guide* McGraw Hill, New York 1940.

WALLANCE, D. *Shaping America's Products* Reinhold, New York 1956.

WEICHER, S. von and GOETZELER, H. *The Siemens Company: Its Historical Role in the Progress of Electrical Engineering 1847–1980* Siemens, Berlin and Munich 1977.

WRAY, L. *The Electrical Appliance Sales Handbook* McGraw Hill, New York 1947.

WRIGHT, G. *Building the Dream* Pantheon, New York 1981.

WRIGHT, L. *Home Fires Burning: the History of Domestic Heating and Cooking* Routledge and Kegan Paul, London 1964.

Articles

ARCHER, L.B. 'Design Analysis 25: Dishwasher' *Design* 163, July 1962.

ARCHER, L.B. 'Electric Food Mixers' *Design* 125, May 1959.

BROOKES, M.J. 'Design Analysis 24: Vacuum Cleaner' *Design* 160, April 1962.

BURNSTEIN, M.L. 'Refrigerator Demand in the US' HARBERGER, A.C. (ed) *The Demand for Durable Goods* Chicago 1960.

CECIL BOOTH, H. 'The Origins of the Vacuum Cleaner' *Newcomen Society Transactions* XV 1935.

COWAN, R.S. 'How the Refrigerator Got its Hum' MACKENZIE, D. and WAJCMAN, J. (eds) *The Social Shaping of Technology* Open University Press, Milton Keynes 1985.

COWAN, R.S. 'The "Industrial Revolution" in the Home: Household Technology and Social Change in the Twentieth-Century' *Technology and Culture* Vol 17 no 1, Jan 1976.

DARLING, C.R. 'Modern Domestic Scientific Appliances' *Journal of the Royal Society of Arts* vol 79 Jan 2 1931, Jan 9 1931 and Jan 16 1931.

De SYLLAS, J. 'Streamform' *Architectural Association Quarterly* April 1969.

– 'The Electrical Industry of Great Britain' British Electrical and Allied Manufacturers' Association 1929.

FORTY, A. 'Electrical Appliances 1900–1960' FAULKNER, T. (ed) *Design 1900–1960* Newcastle upon Tyne 1976.

FORTY, A. 'The Electric Home: A Case Study in the Domestic Revolution of the Inter-War Years' NEWMAN, G. and FORTY, A. (eds) *British Design* Open University Press, Milton Keynes 1975.

– 'Goblin's Golden Jubilee Souvenir' BVC 1951.

GUERNSEY, J.B. 'Scientific Management in the Home' *Outlook* 100, April 13 1912.

HARDEN, L. 'The Design of Domestic Equipment' *Art and Industry* June 1948.

HOWE, J. 'Domestic Equipment – a Survey of Modern Home Appliances' *Design* 103, July 1957.

– 'Interior House Equipment: Illustrated Catalogue of Well Designed Objects' *Architectural Review* Vol 78, no. 469, Dec 1935.

KENNELLY, A.E. 'Electricity in

the Household' *Scribner's Magazine* VII Jan 1890.

MAYERS, J.B. 'Management and the British Domestic Appliances Industry' *Journal of Industrial Economics* XII 1963.

— 'The Market for Electrical Appliances' *Economist* June 24 1939.

NEEDLEMAN, L. 'The Demand for Domestic Appliances' *National Institute Economic Review* no 12 Nov 1960.

— 'Pressed Steel Company Ltd' Pressed Steel 1958.

RICHARDSON, H. 'The New Industries Between the Wars' *Economic History Review* XV 1962.

— 'The Story of Belling' Belling and Co. Ltd 1963.

— 'The Story of Hoover's Sales Policy' *Business* March 1963.

SUDJIC, D. 'Man in a White Box' *Blueprint* 9 July/Aug 1984.

TAYLOR, P. 'Daughters and Mothers–Maids and Mistresses: Domestic Service Between the Wars' CLARKE, J., CRITCHER, C. and JOHNSON (eds) *Working Class Culture* Hutchinson, London 1979.

VAN DOREN, H. 'Streamlining: Fad or Function?' *Design* 10 Oct 1949.

WHEATCROFT, M. 'Industry and Domestic Appliances' *The Manager* Jan 1951.

WILSON, T. 'Electrical Engineering and Electrical Goods Industry' TEW, B. and HENDERSON, R.F. *Studies in Company Finance* London 1959.

WOUDHUYSEN, J. 'Form Follows Fluff' *Design* 416, August 1983.

Exhibition Catalogues

LONDON: The Boilerhouse Project *Art and Industry* 1982.

LONDON: The Boilerhouse Project *Kenneth Grange at the Boilerhouse* 1983.

LONDON: Whitechapel Art Gallery *Household Things* 1920.

ACKNOWLEDGEMENTS

T = top; B = bottom; R = right; L = left

The author and publishers would like to thank the following for their kind permission to reproduce the photographs on the pages listed: Admiral Home Appliances: 20(R); AEG: 10, 33, 59, 60, 66, 78(T), 96(L); BBC Hulton Picture Library: 74(B), 83, 84, 91; Belling: 6, 21, 35, 102; Braun Electric (UK) Ltd: 55, 68, 76(TL); British Library: 20(L), 26, 59; Bulpitt & Sons: title; Chicago Historical Society: 16 (neg. no: 1. W.C.E., 1893); Cona Ltd: 65; The Design Council: 11, 22(B), 23, 39(T), 44(B), 48, 80(L), 81, 100, 96(TR); Electrolux 36; EMI Music Archives: 7; English Electric Co.: 13; Fejer, George: 24; Ferranti Ltd: 53; Fisher Appliances: 12; Frigidaire International: 4; Hoover Plc: 8, 45, 85(B), 103; HMV: 101(T); IAZ International (UK) Ltd: 85(T); Idea Books: 18; International Harvester Co.: 80(R); de Lucchi, Michele: 57; Matsushita Electric Trading Co.: 53(BR); Morphy Richards: 103(L); Moulinex Ltd: 76(BL); National Housewares Manufacturers Assoc.: 28; Panasonic UK Ltd: 105, 97; Philips: 77(R); Prototypes Ltd: Unwin Hyman jacket, 96(BR); Redring Electric Ltd: 43; Royal College of Art: 14, 22(T), 49, 52(T), 52(B), 69, 72, 96, 101; TI Russell Hobbs Ltd: 44(T), 64(L), 68, 72(T); Sanyo Marubeni (UK) Ltd: 61; Trustees of the Science Museum (London): 5, 30(TR), 30(B), 30(TL), 32, 34(B), 34(T), 37, 38, 39(B), 63, 71(B), 74(T), 87; Scott, Douglas: 56; Sears Roebuck: 50; Siemens-Electrogeräte GmbH: 73; Sunbeam: 76-77; Walter Dorwin Teague Assocs., Inc: 31, 51; Tefal: 64(R), 77(BL), 89; Thorn EMI (Kenwood): 75; Women's Electrical Association: 17, 19, 90, 93, 94, 99.

INDEX